ENVIRONMENTAL DATA MANAGEMENT

NATO CONFERENCE SERIES

I Ecology
II Systems Science
III Human Factors
IV Marine Sciences
V Air—Sea Interactions

I ECOLOGY

ENVIRONMENTAL DATA MANAGEMENT

Edited by

Carl H. Oppenheimer

Marine Science Laboratory
University of Texas, Port Aransas

Dorothy Oppenheimer

Environmental & General Reporting
Port Aransas, Texas

and

William B. Brogden

Marine Science Laboratory
University of Texas, Port Aransas

Published in coordination with NATO Scientific Affairs Division

PLENUM PRESS · NEW YORK AND LONDON

Library of Congress Cataloging in Publication Data

Main entry under title:

Environmental data management.

(Nato conference series: Series 1, ecology, vol. 2)
Bibliography: p.
Includes index.
1. Ecology—Documentation. I. Oppenheimer, Carl H. II. Oppenheimer, Dorothy.
III. Brogden, William B. IV. Series.
QH541.15.D6E58   029'.9'5745   76-23295

ISBN-13: 978-1-4615-6926-8   e-ISBN-13: 978-1-4615-6924-4
DOI: 10.1007/978-1-4615-6924-4

Proceedings of the Conference on Environmental Data Management held
in the facilities of the Houston Museum of Natural Science, Houston, Texas,
April 8–11, 1974, sponsored by the NATO Special Program Panel on
Eco-Sciences

© 1976 Plenum Press, New York
Softcover reprint of the hardcover 1st edition 1976

A Division of Plenum Publishing Corporation
227 West 17th Street, New York, N.Y. 10011

PREFACE

Throughout the world a staggering amount of resources have been used to obtain billions of environmental data points. Some, such as meteorological data, have been organized for weather map display where many thousands of data points are synthesized in one compressed map. Most environmental data, however, are still widely scattered and generally not used for a systems approach, but only for the purpose for which they were originally taken. These data are contained in relatively small computer programs, research files, government and industrial reports, etc.

This Conference was called to bring together some of the world's leaders from research centers and government agencies, and others concerned with environmental data management. The purpose of the Conference was to organize discussion on the scope of world environmental data, its present form and documentation, and whether a systematic approach to a total system is feasible now or in the future. This same subject permeated indirectly the Stockholm Conference on the environment, where, although no single recommendation came forth suggesting a consolidated environmental data pool, bank or network, each recommendation indicated that substantial environmental data needed to be obtained or needed to be pooled and analyzed from existing data sources.

I should like to point out that each participant was asked to represent himself only, regardless of his affiliation. The free intercourse of the program was encouraged to obtain a consensus of scientific thought that may provide the environmental data leadership of the future. The strength of the Conference rests on the unequivocable response, free from constraints, either political or otherwise. Therefore, we ask the reader's indulgence to accept the philosophy that the success of this Conference lies in its written contributions in the same free and easy intercourse as taken verbatim during the proceedings.

Dr. Andreas Rannestad

Thirty-four participants representing various types of computer programs and related interests attended the Conference. The NATO representative, the Executive Officer of the Special Program Panel on Eco-Sciences, was Dr. Andreas Rannestad, a solid state physicist from Norway.

The organizers of the Conference, Drs. Carl H. Oppenheimer and William B. Brodgen, are marine ecologists with the University of Texas Marine Science Institute at Port Aransas, Texas, who are currently working on environmental data systems for coastal zone planning.

Dr. Carl H. Oppenheimer

Dr. William B. Brogden (left)

Two groups currently having the capability of interchanging oceanographic data bases are Dr. Thomas Austin, Director, Environmental Data Service, National Oceanic & Atmospheric Administration, USA, and Mr. Dieter Kohnke, Director of Deutsches Ozeanographisches Datenzentrum, Deutsches Hydrographisches Institut, Germany, who are responsible for the United States' and Germany's Oceanographic Data Centers, respectively.

Dr. Thomas Austin

Mr. Dieter Kohnke

The following photographs of Conference participants were taken during the Conference.

(Left to Right) Dr. Robert A. Citron, Dr. James Noel, Mr. Don Rauschuber, M. Lenco, Dr. Russell Eberhart, Miss Diana W. Scott, Mr. Dieter Kohnke, Dr. Wolfgang Kitschler, Dr. Melvin A. Rosenfeld and Dr. John Cutbill.

(Left to Right) Dr. Martin N. Cobb, M. Jean-louis Mauvais, Dr. Shoichi Nambu, Dr. William B. Brogden, Dr. Carl H. Oppenheimer and Dr. James Noel.

Thirteen countries were represented at the Conference. These included the United States, Belgium, Italy, England, Mexico, Germany, France, Spain, Canada, Japan, Luxembourg, Turkey and Scotland.

(Left to Right) Mr. Dieter Kohnke, Dr. Wolfgang Kitschler, Dr. T. V. Loudon, Dr. Robert A. Citron, Dr. Tuncay Saydam, Mr. Juan Toledo M., Dr. Ramon Margalef, Dr. Armand Nassogne, Dr. Hans Jorgen Helms and Dr. John E. Peachey.

(Left to Right) Dr. Armand Nassogne, Dr. Melvin A. Rosenfeld, Dr. John Cutbill, Dr. G. J. Kovacs, Mr. Malcolm Thurgur, Dr. Bruce Lighthart, Dr. J. M. Colebrook, Dr. Dave Berg, Dr. Martin N. Cobb, M. Jean-louis Mauvais, Dr. William Brogden and Dr. Robert A. Citron.

The wide diversity of backgrounds and disciplines contributed to a rich diversity of opinions and ideas.

As the world's populations increase and land and water use change, it is imperative that land and water use criteria or change be documented with the best possible data base before decisions are made. Our environment is so complex that a single group of data or information will no longer suffice for good planning implementation. Before we use any portion of the earth's surface, the question must be asked, "What will this change do to the local area, the local system, the regional area, the continent, or the world?" Smoke introduced into the air has far-reaching significance to the environment. Materials introduced or changes made along our shorelines may influence current patterns and water quality. Changes in upland areas may be significant to erosion and eventual fertilization of the coastal zone or oceanic waters. For example, the fresh water from the Congo River can be detected at flood conditions approximately 1,500 miles into the Atlantic Ocean.

Perhaps, because of sovereign rights, international differences of opinion, or territorial rights of the oceans and atmosphere, the idealist who insists that our environmental data can be pooled and made useful is fighting a loosing battle. However, with escalating costs and an increasing accumulation of environmental data from all sources, it is now the time for idealistic planning. Even if it is impossible to obtain agreement about the types of data needed, the language for data processing, differences in language meaning, etc., the importance of the subject dictates that we must make a start.

Perhaps this concept of need for data coordination is not apparent to the politician who is generally responsible for land use decisions. However, to the environmental scientists who may look at regional or larger environmental balance, the lack of data coordination is a serious handicap and often may put the scientist into a position where he cannot make a value judgement because he cannot describe a discrete environmental balance. The results of projected land use may provide such a wide range of change that it could not be differentiated from natural environmental fluctuation. At such a time, when administrations come to the decision making point and the scientist is indecisive in his environmental analysis, the decision maker may wonder why we have spent so much money to obtain data.

This does not mean that by some magic, when large amounts of data or even all environmental data are in a network for use, that all environmental change can be documented and evaluated. Because at the same time the data base is consolidated a new group of scientists must be trained to utilize and optimize the available data and information in a "systems" approach so that it can be applied to needs for land, air or water use planning.

This Conference was supported in part by the Special Program Panel on Eco-Sciences of the North Atlantic Treaty Organization. Facilities and space were provided by the Houston Museum of Natural Science who generously opened their doors to us; a special thanks to Dr. Tom Pulley and Mr. Carl Aiken for assistance above and beyond the call of duty.

I wish to thank the members of the Conference Organizing Committee, Dr. John E. Peachey, Dr. Hans Jorgen Helms, Mr. J. Heath, Mr. Henry Fleming, Dr. William B. Brogden, and Dr. Andreas Rannestad for their advice and assistance in organizing this Conference. A special thanks to Dr. Andreas Rannestad for his continued support and advice both in organizing the Conference and in editing the proceedings.

I also wish to thank Dr. William Brogden and my wife, Mrs. Dorothy Oppenheimer, for handling all of the details that were necessary for the success of the Conference. And, of course, these Proceedings would not be possible without the expert verbatim reporting of Mrs. Oppenheimer.

I would also like to thank each and every one of the participants for their open and honest comments drawn from their many years of experience. Without this open and honest exchange of ideas and opinions, the Conference could not have been a success.

Finally, I would like to thank the people of the City of Houston for making our short stay in their fair city such a delight.

Dr. Carl H. Oppenheimer
Editor

LIST OF PARTICIPANTS

Dr. Thomas Austin, Director, Environmental Data Service, National Oceanic & Atmospheric Admin., Washington, D.C., USA 20235

Dr. Dave Berg, Stanford Research Institute, Menlo Park, California, USA 94025

Dr. William Brogden, The University of Texas Marine Science Institute, Port Aransas, Texas, USA 78373

Dr. Robert A. Citron, Director, Smithsonian Institution Center for Short-Lived Phenomena, 60 Garden Street, Cambridge, Mass., USA 02138

Dr. Martin N. Cobb, Computing & Applied Statistics Directorate, Environment Canada, Place Vincent Massey, 5 #FLR, Ottawa, Ontario K1A OH3, Canada

Dr. J. M. Colebrook, Institute for Marine Environmental Research Oceanographic Laboratory, 78 Craighall Road, Edinburgh EH6 4RQ, Scotland

Dr. John Cutbill, Sedgwic Museum, Downing Street, Cambridge, CB2 3EQ, England

Dr. Russell Eberhart, Applied Physics Laboratory, John Hopkins University, 8621 Georgia Avenue, Silver Spring, Maryland, USA 20910

Mr. Mike Ellis, EDP Division, Texas Water Development Board, P. O. Box 13087, Capitol Station, Austin, Texas, USA 78701

Mr. Henry Fleming, Gulf Universities Research Consortium, 1611 Tremont, Galveston, Texas, USA 77550

Dr. A. P. Gore, Inst. Terrestrial Ecology, National Environment Research Council, Grange over Lands, Cumbria, England

Dr. James Hammerle, Monitoring & Data Analysis Division, National Air Data Branch, Environmental Protection. Agency, Research Triangle Park, North Carolina, USA 27711

*Dr. Hans Jorgen Helms, NEUCC — The Regional Computing Centre for Education and Research at the Technical University of Denmark, DK 2800 Lyngby, Denmark

---

*Current address—CETIS, CCR, I-21027 Ispra (Varese), Italy

Dr. Wolfgang Kitschler, c/o Bundesminister des Innern, 53 Bonn 7, Rheindorferstr 198, F.R., Germany

Mr. Dieter Kohnke, Director of Deutsches Ozeanographisches Datenzentrum, c/o Deutsches Hydrographisches Institut, D2-000 Hamburg 4, Bernhard-Nocht-Str. 78, Germany (F.R.)

Dr. G. J. Kovacs, Associate Manager, Information Systems Section, Battelle-Columbus Laboratories, 505 King Avenue, Columbus, Ohio, USA 43201

M. Lenco, Ht Comite' de l'Environment, 67B^d Haussmann, Paris 8^c France

Dr. Bruce Lighthart, National Ecological Research Lab., Environmental Protection Agency, 200 S.W. 35th Street, Corvallis, Oregon, USA 97330

Dr. T. V. Loudon, Institute of Geological Sciences, Exhibition Road, London SW7 2DE, England

Dr. Ramon Margalef, Instituto de Investigaciones Pesqueras, Paseo Nacional, Barcelona (3), Spain

M. Jean-louis Mauvais, Ingenieur au BNDO, Centre National Pour L-Exploitation des Oceans, Centre Oceanologique de Bretagne, B. P. 337 — 29273 Brest Cedex, France

Dr. Shoichi Nambu, Head Department of Sanitary Engineering, The Institute of Public Health, 6-1, Shirokanedai 4-chome, Minato-ku, Tokyo 108, Japan

Dr. Armand Nassogne, Commission of the European Communities, Directorate-General "Scientific and Technical Information and Information Management," 29, Rue Aldringen-Luxembourg

Mr. Jack Nelson, Hydro-Biologist, Operations Division, Texas Water Development Board, P. O. Box 13087, Austin, Texas, USA 78701

*Dr. James Noel, NOAA, Environmental Data Service; Dx4 — Systems Intervation and Planning, Washington, D.C., 20007

Dr. Carl H. Oppenheimer, The University of Texas Marine Science Institute, Port Aransas, Texas, USA 78373

Dr. John E. Peachey, Department of Environment, Room N 19/12 2 Marsham Street, London SW1P #3EB, England

---

*Current address—The Toy Factory, Rt. 1, Box 1055, Florence, Oregon, USA 97439

Dr. Andreas Rannestad, Executive Officer, Eco-Sciences Program, Scientific Affairs Division, NATO, B-1110 Brussels, Belgium

Mr. Don Rauschuber, Texas Water Development Board, P. O. Box 13087, Capital Station, Austin, Texas, USA 78701

Dr. Melvin A. Rosenfeld, Manager, Information Processing Center, Woods Hole Oceanographic Institution, Woods Hole, Massachusetts, USA 02543

Prof. Dr. Tuncay Saydam, EDP Manager Marmara Scientific and Industrial Research Institute, Spor Cad. 183/6, Besiktas — Istanbul, Turkey

Miss Diana W. Scott, Institute of Terrestrial Ecology, Monks Wood Experimental Station, Biological Records Centre, Abbots Ripton, Huntingdon PE17 2LS, England

Mr. Malcolm Thurgur, Computing & Applied Statistics Directorate, Environment Canada, 5th Floor, Place Vincent Massey, Ottawa, Ontario K1A OH3, Canada

Mr. Juan Toledo M., Instituto de Biologia, Departamento de Botanica, Apartado Postal 70-k268, National University of Mexico, Mexico 20, D.F. Mexico

CONTENTS

I. INTRODUCTION

Dr. Carl H. Oppenheimer
Dr. Andreas Rannestad

OPPENHEIMER—Good morning ladies and gentlemen. We shall officially open this meeting by asking Dr. Rannestad if he will give us a few words about the North Atlantic Treaty Organization (NATO) who have made it possible for us to meet today to talk about environmental data management. Without any further introduction, I will ask Dr. Rannestad to describe NATO and their interest in the environment.

RANNESTAD—Thank you very much. I will take this opportunity to give you a short outline of the NATO programs in the scientific and environmental fields.

Collaboration and consultation between member countries of the Alliance have been a major concern to the North Atlantic Treaty Organization since it was established. In the mid-fifties, a serious attempt was made to implement NATO collaboration in non-military fields, and a report (Report of the Committee of Three, 1956) from a committee of foreign ministers, Lester B. Pearson (Canada), Gaetano Martino (Italy) and Halvard Lange (Norway), named scientific and technological cooperation as especially important. As a consequence of this report, a position as Science Advisor to the Secretary General of NATO (later changed to Assistant Secretary-General for Scientific and Environmental Affairs) and a Science Committee composed of one highly qualified scientist from each of the member countries of the Alliance was established in 1958. It was the purpose of this Committee to encourage inter- and intra-national participation in environmental affairs.

The NATO Science Programs have changed during the years, but their predominant characteristics have remained an emphasis on cooperation, acting as a catalyst, and providing a capacity for rapid response to new environmental developments. Each of the programs has been conscientiously designed and deliberately implemented to improve an exchange of information between NATO communities. Over 50,000 individuals, of which some thousands come from countries outside the Alliance, including some hundreds from Eastern Europe, having directly participated in these programs. The following programs have been in

1

operation during 1973-74*:

The Senior Scientists Program. This is a small program awarding Science Lectureships, Visiting Professorships, and/or Senior Fellowships to outstanding scientists.

The Science Fellowships Program. This program, administered by the different member countries, awards about 600 NATO Science Fellowships each year. The program has allowed about 10,000 scientists to study for about one year in a foreign country.

The Advanced Study Institutes Program. An ASI is primarily a high-level teaching activity at which a carefully defined subject is treated in considerable depth in a systematic and coherently-structured program. About 50 institutes are supported each year.

The Research Grants Program. The main purpose of this program is to stimulate scientific research carried out in collaboration between scientists in the member countries of the Alliance. Grants are renewable for up to three years and 50 to 100 new grants are awarded each year.

The Science Committee Conference Program. The main purpose of these research evaluation conferences is to identify particularly fruitful areas for future research. The recommendations from the conferences are directed both to those having a responsibility for selecting and supporting research programs and to the Science Committee itself. One or two conferences are held each year.

The Special Science Programs. In addition to the general and more permanent programs listed above, the Science Committee has frequently identified specialized scientific areas as deserving special encouragement or preferential support for limited periods. In 1974 there were special programs on: Air-Sea Interaction, Eco-Sciences, Human Factors, Marine Sciences, Radiometeorology, Stress Corrosion Cracking and Systems Science.

The Science Committee Programs are guided by panels of scientists from the member countries and support is given to all fields of science, with emphasis on fundamental aspects rather than applications. Results from research projects are published in the literature and scientific proceedings from ASI's and conferences are published in most cases.

---

*More information on the NATO Science Programs may be found in the booklet, "Scientific Co-operation in NATO," or the book, "NATO and Science, An Account of the Activities of the NATO Science Committee 1958-72," NATO, Scientific Affairs Division, B-1110 Brussels.

I would also like to mention the Committee on the Challenges of Modern Society which, since 1969, has started and coordinated pilot studies in: Disaster Asistance; Environment and Regional Planning; Road Safety; Air Pollution; Inland Water Pollution; Advanced Health Care; Coastal Water Pollution; Advanced Waste Water Treatment; Urban Transportation; Disposal of Hazardous Substances; Solar Energy and Geothermal Energy.*

In closing, I would like to say a few words about the NATO Eco-Sciences Program which is sponsoring this Conference. The NATO Science Committee, having long been aware of the need for scientific studies in the eco-sciences, started the program in 1971. Under the first chairmanship of Dr. P. McTaggart-Cowan from Canada, followed by Prof. Carl H. Oppenheimer, the Panel surveyed many areas in need of support. One of the areas selected was environmental data systems. The Panel has supported two conferences. The first one was the Use of Electronic Data Processing in Taxonomic Collections held at the Royal Botanic Gardens, Kew, in September, 1973. This Conference on Environmental Data Management is the second one. This environmental subject is very important to all nations and will become even more important as time goes by. I hope that this will be a very successful Conference and that some agreements can be reached.

OPPENHEIMER—Thank you, Dr. Rannestad. I should like to stress that this meeting has been designed to be an informal, round-table discussion, with lead-in speakers. The meeting is designed to discuss Environmental Data Management. The Organizing Committee set up an outline of the meeting and selected Chairmen for each Session who are designed to maintain continuity, and a speaker was selected for each Session to introduce each designated subject. The Conference participants represent a good balance between industry, the environment, theoretical aspects, the government administrative groups, etc. This has been done so that overlapping does not occur and so that we can have a genuine round-table type of discussion. Shall we begin.

---

*More information on the CCMS Pilot Studies may be found in the booklet, "Man's Environment and the Atlantic Alliance," NATO Information Service, B-1110 Brussels.

II. REVIEW OF PRESENT COMPUTERIZED SYSTEMS

Session Chairman
Dr. Melvin A. Rosenfeld
Woods Hole Oceanographic
    Institution, U.S.A.

Principal Speaker
Dr. John E. Peachey
Department of the Environment
England

ROSENFELD—The first topic for discussion is "Review of Present Computerized Systems." Dr. Peachey's talk will introduce this topic.

"Some General Thoughts on Approaches To Environmental Information and Data Management"

Dr. John E. Peachey

Mr. Chairman, it gives my colleagues and I from the United Kingdom great pleasure to be able to join in this Conference. We look forward to learning, to sharing experience and to contributing to the discussion. We congratulate Professor and Mrs. Oppenheimer, Dr. Rannestad and their fellow organizers on all the excellent hospitality, accommodation and facilities provided by their colleagues and the people here in Houston. We are particularly grateful that the NATO Science Committee through its Eco-Sciences Panel agreed to sponsor this meeting and to make possible our attendance, and that it chose such an excellent venue. We recognize that in Houston new ground was broken in data management in bringing together a whole range of interdisciplinary facts and figures for a variety of management applications in support of a concentrated and highly successful program of high technology. It is indeed a great pleasure to be here.

I note the desire of the organizers to keep this meeting as informal as possible, but may I say at this stage that the views expressed in any contribution I make to this Conference are mine and not necessarily those of the United Kingdom Department of the Environment.

We appreciate the willingness of the organizers to re-schedule this Conference to take maximum advantage of the outcome of the series of meetings held in Nairobi which led to a successful conclusion of the Second Session of the Governing Council for the United Nations Environment Program (UNEP).

I mention this because along with the regional consultations that preceded or ran in parallel with the UNEP discussions, the emergence of the UN Environment Program as a potentially viable and across-the-board environmental action plan is itself a major factor in determining in the future the development of capabilities and institutional arrangements to manage and analyze environmental data.

I know the organizers of the Conference want to get down to specifics—we all have to. There is no such thing today as a generalized environmental management information system. If there were, it could survive no more than a few days as a service function if separated from the real life demands and inputs of scientists, managers or policymakers within their own work situations or institutional settings. Hopefully, the future may change this.

We might start by reminding ourselves of the broader framework of environmental management. The inadequacies of our habitat on the one hand and our maladjustments to it on the other are ever present, and we and other organisms we value or study owe diversity and distinctness to this sometimes unsatisfactory evolutionary mechanism. Perhaps the upheaval in the relationships of pure and applied scientists, of managers and administrators, of policymakers and the public, and the general sensitivity as to the appropriateness of particular institutional arrangements may well be symptoms of our increased unwillingness to leave to this mechanism the major role in future planning.

Surely, within the huge resources we possess currently and within the archive, both at the intellectual and systems level, there is sufficient knowledge to pinpoint relevant inadequacies and maladjustments, to assess priorities, and to push forward with promoting solutions in an even more satisfactory manner than before.

It is true that above a certain level of biological satisfaction, the solution to environmental inadequacies today is still largely subjective. Advances in data management alone cannot replace these subjective assessments of priority, but the background identification of real concerns and a balanced actuarial assessment of benefits and risks is always a great step forward. It is well to remember that people often show strange preferences when faced with options concerning the environmental development or protection of the areas in which they live and work. These may be related to lack of awareness or more plainly to economic or more pressing factors, but whatever the cause, a great degree of local responsiveness has to be built into any system of effective environmental management. The supporting data services and research need to echo and fulfill the requirements of this management whilst placing local knowledge about local problems within the broader setting provided by central government and through national and international initiatives.

Scientists, data managers and information specialists are no exception to those who make subjective judgments. They often form strong attachments to the forms of data collection, information analysis, documentation, and publication which were currently fashionable in their formative years. The capacity of these workers to respond to new schemes and to new concerns may thus be initially limited. And so new plans for monitoring or assessing the environment stand a better chance of success if they start from an analysis of what is actually done at present and involve directly those whose present task it is to make routine measurements and who can thus be stimulated into a wider and more contemporary view as to the exciting possibilities of measuring and processing data in more meaningful ways.

As a former biologist, I recently re-examined my own publications—an odd amalgam of a few pieces of relatively good ecological survey padded out with a somewhat embarrassingly long set of reviews which one was constantly being asked to do. This pressure to re-package existing information was particularly noticeable to me during the time I spent in helping to run an intergovernmental information and abstracting service. And in a later job I had a chance to observe the same pressures in the world of information science research and development. In this situation one tried to persuade, stimulate and support information scientists, data managers and librarians to do new experiments to insure that accumulated knowledge could be transmitted in better ways or that intellectual processes could be concentrated at certain key or useful points in the information transfer process. How difficult it is in a complex society to get people to avoid needless duplication of effort in data handling and to commit to technical processing as many of the tasks associated with storage, sorting and retrieval as possible. I look forward, therefore, with a degree of professional interest, to hearing of the progress which data managers are making towards rationalizing their approaches to information provision, notably with regard to ecology and the environment.

It is of interest in this connection that at first sight the chemists seem to have done better than the biologists. But chemists had three enormous advantages. No half-invented Latin terminology was present to produce an artificial binomial "Esperanto" incapable of carrying within two words, however long, any significant part-by-part description of the attributes of the unfortunate species so labeled. Secondly, one has to admit that the chemist, because of this communication clarity, had a closer impact on society than the ecologist and perhaps, therefore, obtained more satisfactory arrangements for data and information provision. Thirdly, chemists were quick to colonize the new field of information science and many of them were well placed to contribute to this important work.

It is true that biologists are now thinking seriously about the general problems of data management and information processing. They recognize that for too long data collections and information services have been developed and promoted in isolation from each other. In all subjects one sees that the multiplicity of uncoordinated efforts remains in spite of considerable advances in the standardization of bibliographic presentation and in the accompanying development of information policy. When one considers non-bibliographic data, the problems are even greater in seeking to standardize a suitable record format. I look forward today to exchanging views on how to overcome this problem.

It is with this background that we approached (during the preparations for the Stockholm Conference, and since in the emergence of the United Nations Environment Program) the way in which we might tackle the total data and information needs necessary to sustain an environmental program at the national, regional, sectoral and global levels. It was decided that, in the United Kingdom, the first step was to establish a national focal point for such matters and this was placed within our own Unit (Central Unit on Environmental Pollution) within the Department of the Environment. We then concentrated on sharing thinking with others deeply involved in the UN Environment Program and in similar regional initiatives. Together we took the first step in constructing a proper management information base for this program and for its national or regional equivalents. It was decided that a simple directory of sources of environmental information was the basic starting point and this led to the adoption and development of the specifications for the International Referral System for Sources of Environmental Information (IRS)* which is now firmly part of the UN Environment Program and for which we in the United Kingdom have started a national counterpart.

The next stage from such a simple system is to begin to establish more precisely defined networks between chosen information or data sources in support of specific environmental management needs. We have taken this step nationally with regard to monitoring and to pollutant information and we are glad that a similar step has been taken by the UNEP Governing Council in its agreement to set up the Global Environmental Monitoring System (GEMS)* which will be based largely on a network of existing or newly planned national, regional, global or sectoral monitoring programs for pollutants, resources, and other environmental factors. A parallel exercise, the International Register of Potentially Toxic Chemicals, is concerned with networking data on the attributes of environmentally active chemical compounds.* To all of these systems we will provide national input and we look forward to continuing to share ideas with our colleagues both here and in other fora on how we can best organize our contributions. (Peachey 1974; Anon, 1974; Anon, 1975).

At this stage here today one can do more than provide a general backcloth against which to learn of more specific approaches in particular work situations.

---

*At the time of publication the most up-to-date UNEP papers relevant to these important global networks and systems are referenced as UNEP/GC/24, UNEP/GC/24/Corr 1, UNEP/GC/25 and the reports of the Second and Third Sessions of the UNEP Governing Council (UNEP/GC/26 and UNEP/GC/55) together with other technical and promotional literature produced by UNEP.

    I think I should elaborate the point I touched on earlier in relation to environmental management. We as a society do accept data and information in a whole variety of not necessarily harmonized ways and from it we draw conclusions which may sustain or alter a program priority or influence a particular decision. We have to take the situation largely as it is and avoid doctrinaire, academic or highly schematic approaches.

    Our main concern must be related, in a line-management fashion, to secure adequate data inputs in order to promote the development and protection of man's environment to the betterment of his immediate and long-term, local and global future, and to the maintenance of those constituents of his biosphere which he seems to think are necessary to his continuing survival.

    So far we environmentalists have produced an appalling weight of documentation, much of it little used. New ways of data presentation, labeling, and greater selectivity must be found to reduce this burden on all of us. The collection of information and data and the production of the resulting documentation is no longer a totally self-justifiable action. We need to address ourselves to the best use to which properly collected and presented information can be put, rather than to preserving the integrity of particular data collections or documents. Sensible analyses of the ways we now tackle data-handling and decide on appropriate formats and computerization give us an opportunity to break the rigidity of the document and to select and deliver specific data at appropriately detailed and not necessarily hierarchical levels for specified management circumstances. Let us now, today, explore these facilities to the best advantage.

DISCUSSION

    ROSENFELD—Dr. Peachey has given us a marvelous philosophical and diplomatic background to our data management problems today. The topic for discussion is a "Review of Present Computerized Systems." I would first, though, like to ask if there is anyone who wants to comment on Dr. Peachey's philosophical background.

    LOUDON—I think Dr. Peachey raised a number of philosophical points on which I would like to comment. He contrasted the nomenclature, or the notation perhaps, by which chemical compounds are described by chemists by their molecular configurations as opposed to the binomial nomenclature of the biologists.

    It seems to me that underlying this distinction is a difference in kind between the model known to chemists in terms of modern atomic theory and the model known to biologists in terms of evolutionary theory. The first is precise, generally agreed, simple, straight-forward; while the latter is not. I think one can see a similar difference between the models of geophysics and the models of the more traditional areas of geology; for example, where an entity can barely be described, where it cannot be precisely defined physically, and the geophysist isn't able to deal with it, it's called geology. The discipline differences in nomenclature can be a reflection of the degree of understanding of the subject, but this does not imply that the nomenclature is at fault.

There is a second related point. Dr. Peachey described a need for common formats for exchange of environmental data. I think there is an important point here. The difficulty in communicating data as sets in the softer sciences, such as biology or geology, is perhaps not the lack of a common format, but something more fundamental than that. A data set itself is not part of the real world. It is an abstraction, which is derived by an individual with all of his prejudices to meet his own objectives in studying this particular part of the real world. He expressed these in terms of a sampling scheme by the particular criteria he used in selecting his observations, particular things he observed, the weightings he placed on different parameters. His look at the real world from a different viewpoint is a perfectly justifiable activity, but to look at another scientist's data set from a different viewpoint can be misleading because one is seeing a filtered view of reality through the wrong filter. One may not see the real world from another point of view, but may be getting a confused picture of someone else's investigation.

So it seems to me that there is something more fundamental than a mere lack of a common format that may prevent full and free exchange of information at a data level between individuals working on different objectives in an area of soft science.

It follows from this that an emphasis on environmental data, as a guide to management of the environment, may be wrong. It may be the scientist's interpretation which has to be communicated, not just his data. I realize that it is possible, within certain well-defined areas, to develop programs with well-defined objectives that result in data collected by a large group of people that can be readily communicated among them. We can see this clearly in meteorology, oceanography, geophysics and, of course, in chemistry. But when you move into other areas such as biology, geology and geochemistry, I think one finds again this difficulty that there is no common agreement on objectivity, and exchange of data out of context can be very misleading.

ROSENFELD—This repeats precisely the argument which I made about six years ago at a Marine Technology Society Meeting (Rosenfeld, 1969). Where Loudon talks about "data in a model," I used the term "data in context."

AUSTIN—Yes, we were there together.

Speaking as a biologist, I can say that some of us are reasonably confident that taxonomists aren't biologists. I think Dr. Peachey has completely overlooked what I like to call the younger generation. I have had the pleasure of exchanging, both nationally and internationally, very large quantities of marine biological data, such as those for the rate of carbon fixation, the standing crop of phytoplankton, the quantity of zooplankton (the forage organisms), or even the number of fishes. This information has allowed a rather new and interesting method of comparing the standing crops and the relative productivity of the Gulf, the Indian Ocean, or other tropical sea areas with the Arctic, or other areas of interest. The important thing, and I think Dr. Loudon implied it, is documentation, which permits whoever uses the data to know how the samples were collected and how they were processed. I think the long years of work of Dr. Colebrook's lab is a superb example of what can be done with biological data per se, as long as you know what you are using. The fact that you have the lumpers and splitters of taxonomy, or for that matter environmental data, should not decry the present computerized biological data whether it happens to be on forests, grains, or oceans.

OPPENHEIMER—I was also prepared to say something somewhat analogous to what Dr. Austin has mentioned about the forgotten younger generation. I believe we have to encourage the development of a new and emerging group of scientists who are willing to devote their time to the management of data rather than the collection of data. How does one allocate the time between the establishment of a data base and the procurement of the original data in the use of a data base?

The concept of data management is to develop a more efficient use and communication of bits of information. This concept can come about, I believe, with the creation of a nucleus of young people who are willing to accept the fact that there are a tremendous amount of data available that need interfacing and collation to establish an interpretive mode. These same scientists may then evolve into a group who will interpret data, which if properly disseminated, can be of use for interpretation and problem solving of today's society's problems. For example, Oceanographic data has had far more unified recognition in the past than coastal data, which is one of the reasons we used as the theme of this Conference, the coastal environment. The coastal zone is a most important area to many nations because it is where most land wastes are identified, where tremendous masses of information have been accumulated and where there exists a need to have environmental data placed into some useful, coordinated mode. Perhaps one of the most useful concepts that we could recommend on the last day of this Conference is the creation of a philosophy of "the" new scientific field of environmental data management.

ROSENFELD—I'm beginning to detect a major division of philosophy coming into our discussion. We are now developing concepts that describe a real difference between data producers, data managers, and data users.

HELMS—There is one aspect of Dr. Peachey's talk that impressed me very much and it may be a little outside your main line of discussion, but I do feel, as a technical computer person, that I should bring it across the table. I observed, during the presentation, that Dr. Peachey made a statement of the present way of disseminating information that implied the use of an old-fashioned style of documentation. This is a problem which we still suffer from very much in computing.

Those of you who produce computer programs will know there is an important aspect of documentation. In a program course you learn certain methods of how to document and probably feel that all these methods are mainly inadequate. I claim that in my field we have never really learned to document what we do on the computers and I also claim that the reason is that we adapt normally old-fashioned methods; namely, methods which we learned in our youth.

Only gradually are we trying to use the computerized schemes to document what we are doing on the computers. There are technical possibilities to do so. The computer language is in a way self-documenting if you use it correctly; if you are willing to give up what you learned in your college days. However, it is very very difficult.

ROSENFELD—Now there is another factor entering our discussions, the differences between young workers and the older.

BERG—I would like to emphasize that several mechanical techniques of computer technology are available for retrieving, for sorting, and for distributing information. Most of the computer techniques needed in the past have been developed and they are continuing to be developed and that is not where the problem lies. The problem of computer usability really lies in the thinking, the management, and direction of getting the right information into general mechanical procedures and processes. Once the information or data is so processed, you can apply the data, so, obviously, the mechanical procedures become improved. But I would like, in review of the present systems, to emphasize that it is important to recognize that a great data resource exists in that area, but there has to be new thinking and new approaches. An example of this is the effort of the National Science Foundation to discover new ways to develop the dissemination of scientific and technical information, and to look at new ways of determining innovation in this area. A great deal of activity is going on in various programs of theirs to show the value of research projects. This will, I think, be of benefit to everyone.

ROSENFELD—That is a very good point, and, by agreeing, we could save much argument during the rest of the week. Is there anyone who disagrees strongly that computer technology is available to do anything we want to do, providing we know what we want to do? Is there any dispute?

OPPENHEIMER—One of the concepts established by the Committee in the first call for this meeting is that we would assume for the purpose of this Conference that the computer technology was available to handle existing environmental information. This basic premise, that we should all accept at least philosophically, was established so that we could emphasize data management and scope rather than computer engineering. An example of our thoughts is perhaps best explained by the fact that the weather of the world is handled as a unit, compressed and synthesized daily from millions of data points.

SAYDAM—I would like to make a remark on the relevancy of data between the point of its initial generation and the point when it reaches the user. When we make measurements from nature we make certain assumptions, one is that there is background noise and therefore some of the relevancy is lost. Then we abstract numerical data from the data and there is another loss. The results are documented and if it has been measured earlier, there comes the bias of the person who is doing the research and documenting and the relevancy may be lost again. I think that although we have very sophisticated computer equipment and techniques and all sorts of complicated displaying of data, the relevancy of the information between the point of its origin and the time it reaches the user has to be considered.

There is another point that I would like to make. If you are not updating environmental data continuously, and that's another point where noise comes, the relevance is further reduced. This is an important facet in data handling.

PEACHEY—The remarks about the advances that have been made in biological data handling are well taken. I think Dr. Loudon's point on the integrity of data is important, but I would have thought there is no difference between the problems of integrity or standing whether they apply to a string of observations or to a document in which they might be published, so this is something that in a sense we have always lived with.

I think we also need to clarify our terminology. To me, data and information are overlapping, if not almost interchangeable terms. I would accept that some data are raw, some are annotated, and some, in fact, could be called mini-documents. And I think we need somebody's help on how we should discuss these different kinds of things.

I would like very much to return to a point which Dr. Helms made. When I first saw a magnetic tape version of a newly computerized data base, I was astonished at the absence of any accompanying documentation on its format, vocabulary and optimal search requirements. The data system seemed to be sterile and I thought the procedures and practices were pretty poor for the bibliographic world when compared to commercial practice in similar scale observations. But, when I look at non-bibliographic data, I am even more appalled at the chaotic way in which data is characterized and formatted.

AUSTIN—I have mixed emotions when involved in a program. If you're the first speaker, you lay the ground rules and scoop what the other speakers are going to say, and if you're the last speaker, you have nothing new to say. Therefore, as to the question about data and information, I think that we must understand what we are discussing this afternoon. Environmental "data," which is the topic, I would define essentially as the digits. "Information" can then be defined as man's interpretation of those digits in published form. I had proposed, and will reiterate this later, that we do not mix "data" and "information." So, if we are going to speak to the data, it might be best if we could speak pretty well to the digits or the numbers. If we are going to speak to information, we should understand it as the published interpretation of the digits.

ROSENFELD—I think that is a very good distinction to keep in mind for the discussions during this week. It is certainly necessary to provide a common basis for mutual understanding for this Conference, therefore, when we talk of data we mean the raw data.

HELMS—Just a small comment. It really goes back to the statement that in this Conference we assume that the technology is available to do everything we want. I want to put in a small warning about that statement. I won't elaborate on it today; I will probably do it later in the week. However, Dr. Peachey's example with magnetic tape is an excellent example to show how inadequate we are in our computing systems; how far we still have to go before we have systems which can really work as we want them to work.

To supplement Dr. Peachey's excellent example, let me tell you that I run just a general-purpose computing establishment for research and education. In all such establishments you have to employ detectives to trace and use the magnetic tapes which users bring in and which they receive from colleagues elsewhere. Nowadays we must use a specialist just to do that. What a waste of time! What a splendid example of the inadequacies in our present technology!

ROSENFELD—I agree with you because I also run a center and we get tapes from all over; your warning is well taken. Whether the question of incompetence in labeling tapes falls under what we are going to accept as technical capability, I don't know.

KOVACS—Dr. Peachey, I believe you made reference to an on-going program in the United Kingdom to put together a directory on environmental information and sources. I have a problem with the concept of what environmental information is. For example, consider an environmental impact study, one that is to analyze the impact of constructing a nuclear power facility in a certain region of this country. The information you have to draw on comes from virtually all areas—social sciences, physical sciences, engineering, biological sciences, etc. Consequently, a directory of environmental information services might, and perhaps should, encompass an extremely broad category of data.

OPPENHEIMER—Before Dr. Peachey replies I should like to emphasize the comment of Dr. Austin on the difference between data and information. I believe, Dr. Kovacs, your statement relates only to information.

PEACHEY—Well, we started with the overall national response we had to make to UNEP in terms of the information exchange necessary to sustain the program. Every global system, like the original plans for UNISIST, starts off as a -system, becomes a network, and is finally regarded as a set of agreed procedures. And jolly good procedures they are, and useful, but you get this process. So with people here and elsewhere we tried to identify which level of environmental data management would be appropriate for UNEP to aim at, bearing in mind its special status as a coordinating body rather than as a full-fledged operational U.N. agency. It seemed appropriate to start with a comparatively low level of information exchange—somewhere between the "yellow pages" and the "switching center" concepts. This led to the formation of the International Referral System (IRS) for Sources of Environment Information to which we and colleagues in many European countries and over here have agreed to contribute. Now, will it also become a useful general systems design and format for more detailed inventory-taking tasks? We are trying to persuade UNEP that this could be so and that it should use IRS procedures in all its inventory-taking tasks.

And so I think the problem of coverage is that the data file grows naturally in response to need. Once it gets above a certain size you can play games so that the system becomes a quarry for identifying people/subject connectivity. Hence, you can really define subject interest in terms of expressed concern and extent of relevant activity.

OPPENHEIMER—I feel an undercurrent here that I would like to develop and perhaps sow some seeds of thought for the rest of the Conference. That is that the ultimate goal of an environmental data system would be a system that would allow me, as a scientist-user, to go to Mr. Berg and say, I've get a terminal I would like to connect up to your data to compare your information with mine; I'd like to approach BASIS to see how our information compares. I should like to go to the Chesapeake Bay information data system, and from my terminal, ask questions so that I can compare our Texas Bays with his very large estuary in Maryland. I think that if we use this concept and ideology as a sort of a common denominator for a goal, we can discuss the ability of the information user to access comparative environmental data from other areas.

EBERHART—Dr. Oppenheimer commented on scientific aspects of the utilization of environmental data bases. I would like to bring up something that we are very interested in the Chesapeake Bay Area, and that is a management use of our environmental data. We consider two users of our system, both research and management; and we are very interested in trying to understand and predict environmental, economic and social consequences of various alternative management strategies which might be employed by different Bay Area regulatory agencies on the Chesapeake Bay. This is one of the primary interests of agencies like the Virginia Institute of Marine Science and the Maryland Department of Natural Resources. We want to provide scientific use of the data, but we also want the data to be in a form which can be of use to management agencies.

AUSTIN—The Chesapeake Bay in our part of the country is known as the Virginia Sea. An interesting development is taking place in terms of environmental data. Probably a lot of you can come up with examples, but I will try to use that of wheat yield. Through advances in technology and various types of fertilizers, the midwestern United States wheat yield is now about optimum at 30 bushels an-acre. An appreciable increase in that yield is going to require a significant, or quantum jump in technology. As a result of situations such as this, environmental data are now becoming, relatively speaking, very important because of the already advanced sophistication of technology. There was a time when a drought, a flood, a cold spell, or a warm spell had less of a significant influence, because technology was bringing the improvements. But now, we have to begin thinking more and more about the significance of environmental data. A drop from 30 bushels to 27 bushels per acre is a very significant decrease in yield or production when we consider the whole middle west of the U.S. Such a decrease now can result from a drought, a flood, or other problems during the germinating season, or the impact of weather on harvesting. We have got to find a way to use environmental data more quickly, and more cost effectively to relate to such national changes.

ROSENFELD—That is quite true and introduces another division we can make in our theme. We should distinguish between need for speed for synoptic or early use of the data and the more relaxed requirements for research needs. We must be aware of the cost differences, which are large and which are perhaps not justified for normal ongoing research, as they are for synoptic needs.

OPPENHEIMER—Again I bring to your attention the advances of the weather bureau in the utilization of millions of data points to produce weather information.

PEACHEY—Somewhere between Dr. Helms, myself, and the silent Cutbill, there is an idea hovering here which I think needs putting on your list. It is something to do with the real difficulties of data management when a multitude of dissimilarly organized bureaucracies are involved in a hierarchical manner. I feel particularly conscious of this since the U. K. joined the European Economic Community (EEC). Imagine having a whole new layer of integrative data management superimposed on what once used to be the end of the line. You must have this in the U. S. Federal structure too.

HAMMERLE—I wanted to make a comment on this very same subject. For example, in EPA there is a directory of environmental data systems that covers 20 pages. These are not all totally environmental data systems, but within EPA there are that many systems of which many are devoted entirely to environmental data.

We have just completed a survey among state and local agencies in the United States, newspapers, industrial organizations, research organizations, and also with Battelle and Stanford Research, to ascertain who may want direct access to our data systems. And it turns out that they don't, in general, because everybody wants to have his own thing. They are confused between data and information. They want, however, to have the data that they use most frequently in the form that they use most frequently. And so, therefore, I don't think we will ever see everything all tied together into one gigantic system.

At the same time there has to be some common denominator which permits people to use data from other systems and to put it directly into their own system. So I think the concept is confusing, at least from what I thought of a year or two ago. I thought that we could have one large system which would tie everyone together. But it turns out that people, at least this generation which is involved with data, don't want that. They want their own thing. And we have got to have standards. We have got to have standardization.

It wasn't clear from your previous comments whether you were in favor of it, but that's the key to all of this. Standard methods of measurement and analysis must at least be identified so that we all know what we are talking about; standard identifiers for geographical areas or pollutants or whatever we are talking about; standards for system documentations; standards for systems operations; and standards to define the quality assurances associated with the data. All of these things have to be accomplished.

There will be a development period during which everyone will continue doing their own thing while these standards are developed. This development period will be followed by a period during which everybody uses those standards and at the same time goes back and converts existing data systems to those new standards and also converts non-computerized data systems to those standards. So there is going to be a long period of time in which a gross amount of confusion will exist while standardization concepts are developed. Then after they are developed there will be a long period before everybody gets on to that boat and moves forward all together.

And there is going to be a tremendous amount of conflict; I see it every day in my work because each individual state and local agency wants to do his own thing. For example, within EPA the organizations that have different responsibilities want to do their own work completely independent of each other. And it's going to be a long educational process. As I was mentioning last night, a super salesman is going to have to get involved here to sell this concept because, although we may agree on this, the people who do this every day with their daily work don't accept this.

OPPENHEIMER—Perhaps the difficulties you mention are due to the differences in opinion between data and information.

LOUDON—I find myself somewhat in conflict with some of the undercurrents Hammerle has been mentioning. I wonder if I can go back a little bit to the point where Dr. Austin drew the distinction between data as being digits and information being a matter of interpretation, in published form, of this material. I have no trouble in accepting this. But he went on to suggest that these two things must be kept separate, linked, no doubt, but kept separate in the way they are held and the way in which they are presented. It's that latter point with which I take exception. My feeling is that it is necessary to integrate these two aspects in the computer. An example that comes to mind is the work which is going on at the Institute of Terrestrial Ecology (ITE) at Merlewood, which Mr. Gore is associated with, where their use of the computer is largely in the field of modeling. The starting point of an investigation is presumably the relationships that they wish to determine. They attempt to represent the relationships as a computer model and collect data in order to test this model. This perhaps offers one way in which another user can come in and get at that data—through the model. I don't see myself, as an outsider, being able to get useful access to information about, say, an estuary without having access to the model for which the data have been collected. And I would seriously doubt whether it is possible to get sufficient uniformity. and standards to cover the whole range of environmental data. I don't see how it can be done.

It was perhaps because of my conflict with that particular undercurrent that I find myself in conflict with the other one which is that the technology is adequate to cope with what we want to do. Because for me, it isn't. The concern of the environmental scientist is very often with "fuzzy" data sets. The geologist has the feeling that this sandstone was deposited from mountains which were over here somewhere, or maybe over there, but almost certainly not over there. How do you express this in a satisfactory way with a computer? I just don't know. I could pick up a piece of sandstone, I could look at it, I could attempt to describe it to another geologist in intricate detail using a computer method or simply writing on a piece of paper and the geologist might be totally unable to identify the age or locality of the sandstone from the description. If I handed the specimen to him, he might identify it immediately. The technology, or perhaps the notation is inadequate. I am not able to express imprecise concepts adequately through a computer. So I find myself somewhat in conflict with your undercurrents.

ROSENFELD—The conflict is what makes the undercurrents. I think we could spend a great deal of time arguing this issue and perhaps we had better clarify it since we have two days ahead of us. I believe we were talking about the electronic and communication hardware aspects of the computers and the ability to have systems software too. Now you are really talking about the adequacy of the geological model.

LOUDON—No sir, I'm talking about the adequacy of the software to represent the geological model.

ROSENFELD—In that case, we will have to keep this open for continuing discussion.

EBERHART—I would like to respond to some of Dr. Hammerle's remarks on standards. From my own parochial point of view I feel that one of the biggest traps we could fall into would be the imposition of standards on users, and I think that the success or failure in many cases of these data systems will depend upon their flexibility—the ability of these systems to respond to individual users who are doing their own thing. I think that this is a necessary requirement of a data system, and that they are going to fail if they are not able to respond to these people who are doing their own thing.

HAMMERLE—There are two concepts—those people who collect data and those people who manage the data that the collectors have provided to them. And I guess I was addressing primarily those who collect the data and keep it as opposed to those who use it. I wasn't indicating at all to restrain the way that people use data, that certainly ought to be left open at all times.

ROSENFELD—The distinction is perhaps between a given model for collection, as compared to the shotgun approach of taking BT's (bathythermographs) all the time anywhere over the ocean where they are recorded and then set away for future use. What I think Dr. Loudon was trying to distinguish was between a design model to answer questions and a more general collection model.

FLEMING—I just wanted to comment in regard to Dr. Hammerle's statement. I rather agree that we have to let the investigator choose whatever system he wants to use but I think you covered yourself when you said that he has to define what his operations are and define his system. The difficulty is that we don't know how the people collected the data and we don't know their program parameters.

AUSTIN—Yes, I think we are now moving into the area of data management which is now the name of the game: I don't agree totally with Dr. Loudon. We have what we politely call the primary users, the people who have collected the data for a purpose. These data are next brought into the appropriate center. Regarding the BT's you mentioned, we now have digitized about 780,000 from the oceans of the world. These processed BT's allow us to look at the changes in our oceans since the BT was first invented in 1938. The data were collected for a variety of purposes, including research at fixed stations, the ocean weather ships. The U.K., the Scandanavian countries, and others, require their ships to take BT's no matter where the ship is going. These data for the secondary users are, I think because of the nature of my job, more important than data gathered for the primary users. Last year, if you consider information and data in the definition I propose, we responded to 110,000 requests from users. Incidentally, there seems to be an antipathy here to the use of the word "users." I think it is a wonderful word—it describes exactly why we manage data.

ROSENFELD—Now we have another concept appearing: one, where

strictly modeled conditions in obtaining data locally answer a specific model question, which is what Dr. Loudon was talking about; and two, the use of masses of data that were collected either under another question—asking model—or under routine conditions as described by Dr. Austin.

LOUDON—Can I clarify a little the point that I was trying to make. To take an oceanographic example, one might consider the bathymetry as presented on a chart as a set of points collected by various ships. If I were a user of the data, a secondary user, what I would like to be able to do would be to apply a contouring algorithm supplied by the primary user, because he knew how the data had been collected and he felt this was the right algorithm to apply to his data. I would feel unhappy if I used my own algorithm because I might be using one which was inappropriate for the data collection technique.

AUSTIN—To take that specific example, if you worry about catching up the backlog of data, each nation would probably use up a very significant per cent of its resources. So you have to make certain onerous decisions. One we made for bathymetric data is that unless the user really wants it, we do not provide him with bathymetric data that have not been collected together with precision echosoundings and satellite navigation. This means we are developing a high quality base that my great grandchildren could use with a degree of confidence that they would not have in data collected along with lead line soundings and dead reckoning position fixing after a three-week period of overcast skies. So I think our data management techniques must be changed as technology becomes more and more sophisticated.

ROSENFELD—Quality control is important.

OPPENHEIMER—I would like to pull together a few of the thoughts of Dr. Hammerle and Tommy (Austin) to make a point. We should not, from an ecological standpoint, overemphasize the dichotomy of separate information systems. It has been our experience that with sufficient data points one can start to identify certain cluster characteristics and point out possible anomalies of poor data points. If we could consolidate all of the estuarine temperatures of our coastal regions in a manner that has been done with BT's (bathytermographs) in the open oceans throughout the years, we would be in a far better position today to assess diurnal and annual changes and thermal pollution. Myriads of little bits of data do exist for temperature, proprietary in some cases, but in little individual systems which we might call mini data banks for lack of a better word. And I think that a system that evolves for unified environmental information should not have any restriction on the merger of data bases. From my own experience, from what Bill (Brogden) and I have been doing, and from what the State of Texas has been doing, I feel that the more data you can get into your system, the better chance you have of coming up with some rather good information that allows analogy and interpretation.

RANNESTAD—I do not think that we will ever go so far with standardization that all scientists will use a standard way in collecting, recording and computerizing data. The main task of this Conference, however, is to develop a concept for the management of available and future data systems which will overcome present deficiencies in such a way that data can be retrieved in a manageable form, and to achieve this it is necessary that data banks are inter-operable and can be interconnected.

A minimum requirement for inter-operability of data banks is a standard description of the data banks and the data, and that this description be included on the front of each data tape. This will solve some of the problems, but full inter-operability can only be reached if standard descriptors are used for the data. I do not think that it would be possible to come to an agreement for a complete set of descriptors, but a minimum set of standard descriptors would be a big step in the right direction.

I do not agree with the statements that the present computers are inadequate for the task at hand. The present day computers are, in my opinion, adequate to handle any reasonable problem that you may assign to them. The inadequacy is owing to the limitation in the programs available and, even more, to the inability of the operators/scientists to give the proper description or ask the computer the right questions in the right way.

OPPENHEIMER—I thoroughly agree. We must, as environmentalists, approach environmental problems with the philosophy that basic data assimilated in coordinated data banks can be interpreted or disputed at will. However, the existence of a general data base implies that users must be consistent with general ecological concepts. Perhaps this fact is one of the most opposing concepts to a general data base. Users will have to support their issues from a common data base.

I believe the accumulation of data will never be a waste of money as such data or baseline information can be the basis for all environmental decisions in the future.

PEACHEY—May I add this comment. I think the data concept is something which may be a hazard to us if we don't sort it out. I think some of us have the feeling that there is probably a fundamental difference in the accountability of environmental data managers on one side of the Atlantic from the other.

I think that in Europe we have swung round almost without knowing it to a very goal-oriented situation. It used to be the other way around, it was the U.S. who used to talk about mission-oriented effort and we took the lofty, thoroughly New England, detached line. But I think something has happened and I think that there is going to be a problem in understanding the way we talk with each other unless we get it clear. And it seems to me there has been a tremendous swing in certain countries toward measurement in direct support of known policies. This is not to undervalue the role of what we hesitatingly call the archive.

It seems to me that, and I might be corrected if I am wrong, in the States you have had a tradition of data librarianship (and I don't use that term abusively like I normally do) which on our side of the Atlantic we never had.

I think it is a pity we haven't the spacemen here to tell us how they brought stuff together in a condensed form and then expanded it when something went wrong to find out what went wrong and then condensed it again. I think this is a very useful and relevant analog for environmental data management. You have been at this long enough to develop a library approach. In other words, not only in direct support of problems, but also speculatively, as should perhaps be. But I think that some of us in Europe have swung, perhaps almost too far, away from building the archive.

ROSENFELD—I would say, and I think probably Tom (Austin) would agree with me, the archives didn't start until sometime after data collection. The data were never collected specifically for the purpose of the archives. The archives started when the government realized that it was spending so much money on acquiring these data and they were being dissipated and becoming unavailable after a number of years. You could not find the man who had collected them, you could not find the records, etc. And then the archives started. Is that correct?

AUSTIN—Reasonably so. The initiative in this was, of course, the IGY—International Geophysical Year—when man began to realize that he was making tremendous investments in the collection of environmental data. The World Data Center system concept was a post-IGY concept. The National Oceanographic Data Center in the United States opened its doors in 1960, The National Climatic Center opened its doors post war, 1951. There has been a swing of the pendulum. Dr. Burcas is the Director of the Oceanographic Data Center in his country, as you know. There is a BODS, British Oceanographic Data Center in Wormley, U.K., another in France, an excellent one in the Scandanavian countries, and one in Spain. There are now data centers in 24 different nations. With them has come the realization that environmental data are a national resource to be husbanded, that they are a real currency for international exchange. Governments are recognizing that they must protect their national resources if they are to understand what is happening in their nation and to its environment.

BERG—I have a question. I wonder if you could give one or two concrete examples of what you were referring to, Dr.Peachey.

PEACHEY—The first thing that happens to you when you leave the research bench and become a government interpreter is that your examples no longer sound real. But some of us just took stock after this morning and I think first of all we have a very genuine difficulty in interpreting the line management structure, if I may be frank, in your government's center. We have research organizations and my colleagues from the U.K. Research Councils will all explain their work. What we don't have, in general, are agencies with delegated regulatory functions, so it is very difficult for us to understand where your management sits on top of your data service, and by that I mean management to enforce or identify environmental quality standards, legislation and that sort of thing. All we are saying really is do bear in mind if we appear rather sort of nasty, practical people, it is because we now feel under this direct line management constraint, whereas, we feel that you in the U.S. possibly are just a bit more relaxed, that you can say we have this general service function, you can take it or leave it, but here it is.

NASSOGNE—You see, I think it is important to see what definition you use for "the total system" because there are two approaches to this program—you have a set of existing data, you want to put it into a computer form and this is to a system which has access to all kinds of data; or you try the other approach, which is to define first what is your total system, what you are expecting from the system, what kind of data you think important to put in the system after you see the coordination.

ROSENFELD—It is my job as chairman of the first session to attempt a summary of what we have discussed this morning. As far as I could detect, our discussion was wholly philosophical and, probably, has laid the groundwork for the rest of the week. There were raised a large number of questions and the beginnings of important conflicts were stated; there were no answers. There was nothing in terms of conflicts, undercurrents or differences that have not been expressed many times over in other meetings. This is not bad, it simply means that the old problems are still with us. Our aim for this week is to come up with a recommendation for a "total system." We all realize that if we can do this, it, will not wipe out long-standing conceptual differences. If we are successful it will be because we learn, this week, how to accommodate various apparently incompatible views. It is safe to predict now that any "total system" will have multiple facets.

It is not possible this morning, despite attempts by your Chairman, to even get provisional and temporary agreement on two rather conceptually simple ideas. It was thought that, if we could agree on these (even if they were not rigorously true), there might be two items we would not have to argue about during the week. These were:

1. The idea that, with respect to computer technology (hardware and software systems), we have the power, or know that the power can be developed, to achieve any system we are intelligent enough to think up and that we are wise

enough to describe what is necessary to do.

Many of the participants apparently will accept this. Dr. Loudon, however, is doubtful about the adequacy of software to represent a geological model. Also, Dr. Helms warns about our inadequacy in technology. It seems to me that his warning is, possibly, more about personal incompetence than about our present technology.

2. The idea of Dr. Austin that "data" and "information" have two distinct meanings and should not be confused or interchanged. There was even some discussion that these should be kept separate and treated separately. Dr. Loudon agrees with the definitions but takes exception to the separation of data from the accompanying information.

Probably the two most important ideas expressed this morning are:

3. Dr. Peachey's concept that the data and information are for real-life purposes (e.g., to manage an environmental system). The importance of line management responsibility in contrast to librarianship (archiving) was stressed.

4. Dr. Loudon's repeated insistence that the starting point (model) for the collection of data is of the utmost importance to any subsequent user. To me, and I have said this many times over the past few years, this factor cannot be over-stressed. This morning there were many comments on this topic; clearly any ultimate system will have to provide a secondary user with much more than raw data.

Other points of discussion in this session covered a wide range of items relevant to an ultimate system. Perhaps these will be aired further during the week. They are presented here in no particular order of importance:

5. The question of standards. By "standardization" do we consider all the levels at which that word can be applied from experimental methodology, collection methodology, formatting, data entry into a system, manipulations, system designs or simply standardization of the method or technique by which anyone can get into dissimilar systems? We will have to "standardize" or delimit the use of the word "standardization" to specific reference to a particular level or levels of action.

6. Quality control. Again, this is closely related to the information about the data. Not all data-base managers are as careful as Dr. Austin is about the bathymetry data. However, it is still possible for users to work with data of less than superior quality if the whole of the information about the data is made available to them. Let the user determine the quality of the data, and let him understand the model under which they were collected.

7. There was some discussion about the different attitudes and capabilities of younger workers and older workers. Many new ideas and concepts will have to be done by clearing out old ideas; I hope this doesn't necessarily mean the clearing out of the older worker. Dr. Oppenheimer talks of encouraging the development of a group of younger people into a new discipline of data management.

8. Are there different kinds of data according to the model in which they were collected or to the purposes to which they are to be put? This morning we began to make distinctions between data collected in the framework of a scientific research model and data collected routinely on "ships of opportunity." We also can distinguish between data needed synoptically for immediate use (e.g., weather forecasting) and data for leisurely research analysis.

9. The question of what are "environmental data" will have to be determined during this week.

10. Finally, I know that we will be occupied during the week with discussion of what we mean by a "system." It is hopefully agreed now that this does not mean one gigantic, unified, centralized, standardized system. Perhaps it means as little as the ability of a researcher to obtain data and information rapidly, in some minimally standardized manner, from a well-indexed set of widely dispersed data bases.

One topic not even mentioned in this session was cost. Clearly, any approach will have to consider cost versus benefit or value. The more generalized a system is, the more it will cost. Careful analysis will be required to determine what share of available resources should be apportioned to a "system" and what share to individual users.

SCOTT—I would just like to say one thing that struck me rather strongly this morning; you were distinguishing between the people who collect data, the people who manage it, and the people who use it. Now, I think that this is a rather dangerous thing to do. I think that a data bank of any sort must have a link between these three people. In other words, if you are collecting data just for the sake of it, then you are wasting time and money. Unless you have an idea when you start of how it is going to be used, when you come to use it you are either going to find that you haven't collected something you needed to collect, or else that you have collected a lot of redundant information. I think that a lot of thought has to go into this. A data bank must have a specific purpose in mind if it is going to be worth the money that is going to be spent on it.

ROSENFELD—Fine, that was discussed this morning. I'm sorry that was left out of the summary; it was, in some sense, assumed under the model that Dr. Loudon was speaking of. But I think that is a very important point.

I believe we have brought out some areas that will require further discussion during this Conference and that in itself has made this a fruitful session. On this note I will terminate this session with thanks to the participants for their lively discussions.

III. EXISTING BODIES OF DATA—NON-COMPUTERIZED

Session Chairman
Dr. John Cutbill
Sedgwic Museum
England

Principal Speaker
Dr. Thomas Austin
National Oceanic & Atmospheric
Administration, U.S.A.

CUTBILL—The theme for this afternoon's session is "Existing Bodies of Data—Non-Computerized." I take it that that includes on the one hand all the files we have put away in archives somewhere, and on the other hand all those active bodies of data that have been taken and used which people don't have enough money to computerize or just really don't want to computerize. What there is to discuss in that, I believe, will come out of this afternoon.

I want to throw in two questions before we get on to the main contribution. The first is that we have been trying to lay to rest the adequacy of our data management techniques. We may be quite right but there is a very simple problem, we can't afford to use some of our more extensive systems. There is a very real problem of support particularly for those who have archival data banks. Dr. Loudon, of course, has just this problem. He is Liaison Officer for the National Geological Data Bank for Great Britain and somebody's lost the key. And the question I want to ask is what are the guidelines for the use of this mass of data that the policymakers, both the people that are immediately responsible and the next level of government up, should be getting at in this area to decide what investment to make into mobilizing that area? That is one question. The other question perhaps more directly to this is: are there any real differences in managing these kinds of data? Is there anything fundamentally different about a data base that is computerized and one that is not computerized? Is it merely a matter of eco-technique or is there really something of significance involving the capability of accessing a computer program vs. traveling across the country or the globe to inspect the myriad of environmental data files that are scattered throughout the scientific community? These are just a couple of starters. Dr. Austin is now going to have a chance to put forth his views on this.

"Existing Bodies of Data—Non-computerized"

Dr. Thomas Austin

When I began to develop this talk, paying attention to the suggested title, "Existing Bodies of Data—Non-computerized," I realized that I spent most of my waking hours, along with the people with whom I work, reducing existing bodies of non-computerized data into computer-compatible form. It wasn't an easy topic to work with. Mr. Noel, who spoke to you this morning, with whom I work, has told me two or three times in the last week he's waiting to see how I'm going to avoid discussing computerized data bases. I'm not going to. There is no way that one can readily restrict comments to existing bodies of non-computerized data. However, I would like to, as I mentioned to you this morning, speak to existing bodies of files, data, and information that cannot in their present form be manipulated by ADP or EDP equipment. By no means does this suggest that the data we will look at (whether it happens to be analog or in logbooks) cannot be converted into computer-compatible form. Of course it can. All you need are the resources.

So we'll speak about data that are not on punch cards, punch-paper tape, magnetic tape, disk, drum, or what-have-you. This morning I mentioned data and information: data are the digits—the numbers as recorded or manipulated. Rather loosely defined, information is the hard copy—the journals, books and abstracts, man communicating the results of his experience, observations, and research to inform his fellowman. The point came up this morning that a user who wants the data also wants the related information. We in NOAA (National Oceanic and Atmospheric Admin.) provide him with author, subject, abstract, and journal reference. Thus, he is less apt to plan an experiment to reinvent the wheel. He builds on an existing fund of knowledge.

I propose to organize what follows (because I'm primarily going to give you examples) around environmental data, those relating to the atmosphere, the oceans, and the solid earth. These are the data that we in EDS (Environmental Data Service) are involved with each and every day. I propose to further restrict my discussions to those data in three of EDS' national data centers: the National Oceanographic Data Center, the National Climatic Center, and the National Geophysical and Solar-Terrestrial Data Center (Figure 1). There are two other centers in EDS: the Environmental Science Information Center—this speaks to libraries, subject, author, and abstract that I have indicated I will exclude—and the Center for Experiment Design and Data Analysis. This Center emphasizes that data management begins with the concept of the experiment and that the data manager is involved from the concept through the design of the experiment, testing of sensors, operational phase, making the data available to the primary user, and then making the data available to the secondary users.

Also, I would speak briefly on the World Data Center system, which evolved as a result of the International Geophysical Year. World Data Centers A for each of the disciplines oceanography, upper mantle, glaciology, geology, geophysics—are in the United States. World Data Centers B are in Moscow.

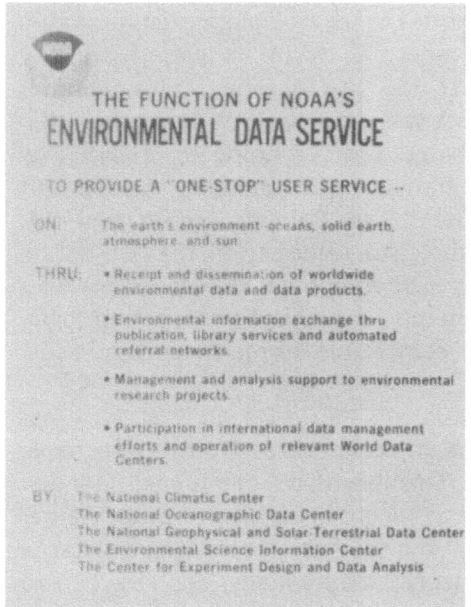

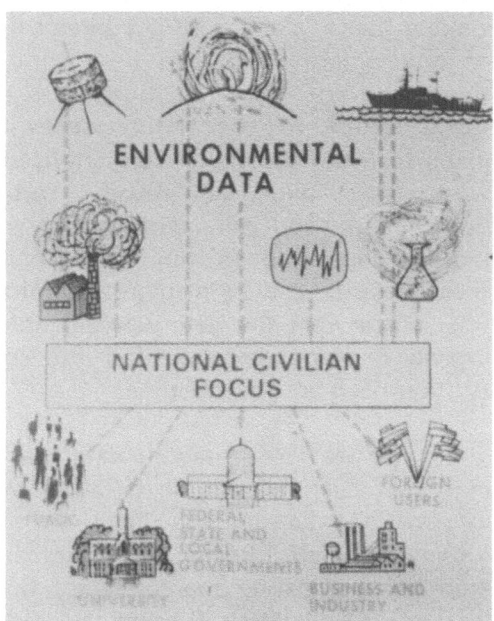

FIGURE 1.   The function of NOAA'S   FIGURE 2.   Scope of NOAA data.
             Environmental Data Service.

Figure 2 diagrammatically shows you the different sources of the data which move into the Centers, from both national and international sources. Across the bottom of the figure is the spectrum of users. Although the Environmental Data Service is within NOAA, that organization makes up only about 18 per cent of our total user community. Of the 110,000 requests to which we responded last year (66,000 were for environmental data/data products, the remainder for information), 82 per cent were from other Federal agencies as well as state and local organizations within our nation, industry, scientists, and for international exchange.

I shall briefly discuss data and information collection and the formats used to record them. When man began to observe phenomena about him he quantified to the base 10 because of the number of fingers and toes. Then he began recording—tying knots in ropes and making piles of stones, to keep track of things. He developed simple instruments to extend his senses. In communicating with his fellowman, he prepared analogs (Figure 3). As the sophistication of

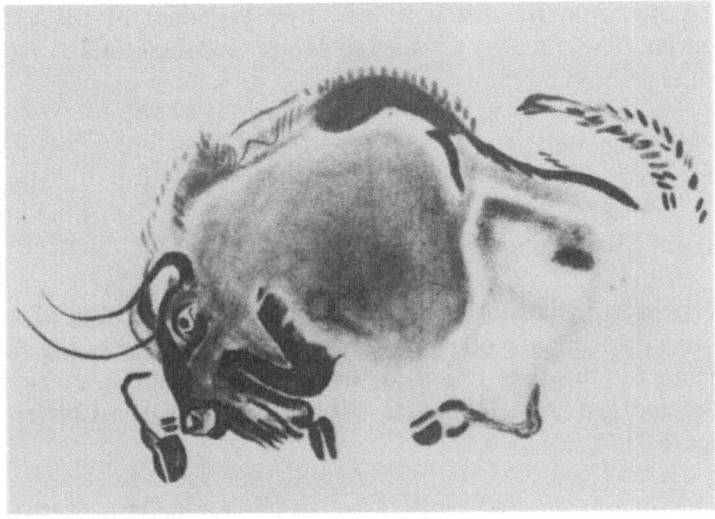

FIGURE 3. Man's earlier attempts at recording.

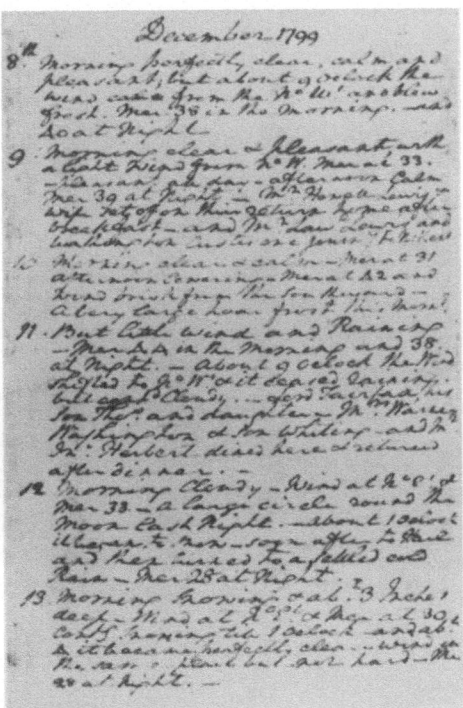

FIGURE 4. A page of weather observations recorded by George Washington.

instruments progressed, he read them and recorded their measurements. Figure 4 is a weather record from the diary of the first President of the United States, George Washington. This record, an analog, is not a computerized record by any stretch of the imagination.

Man communicated data in other ways. Ben Franklin, who collected the logs from ships that sailed between the United States and Europe, noticed a northerly set in course as recorded in the ships' logs. He identified this effect as caused by the Gulf Stream and prepared a chart (Figure 5).

Round-the-world transportation opened up the possibilities of two or more men working together on a global scale. An early example is the cruise of the Challenger (Figure 6), the first round-the-world oceanographic cruise, which took place 100 years ago last year. The data in the logs from the Challenger are still being used (Figure 7).

FIGURE 5. Ben Franklin's chart of the Gulf Stream.

FIGURE 6. Research Vessel HMS Challenger.

Number of Sounding	Distinguishing Number of Station	Date. 1873.	Latitude.	Longitude.	Depth in Fathoms.	Nature of Bottom.	Temperature of the Sea-water. Bottom	Surface	Specific Gravity of Sea-water at 60° F. Distilled Water at 39°=1. Bottom	Surface.	Trawling or Dredging.	Serial temperature stations marked	Place to which the Station is shown.
			NORTH.	WEST.			°	°					
41	VIIr	February 11	28 42 0	17 8 0	1750	Volcanic mud.	37·5	63·0	...	...	...		5
42	VIIu	,, 11	28 20 0	17 34 0	1340	Volcanic mud.	38·5	65·0	...	...	...		5
43	VIIv	,, 11	27 58 0	17 39 0	1620	Volcanic mud.	37·5	65·0	...	...	Dredged.		5
44	VIII	,, 12	28 3 15	17 27 0	620	Volcanic mud.	...	64·5	...	...			5
45	1	,, 15	27 24 0	16 55 0	1890	Globigerina ooze.	36·8	64·5	1·02650	1·02730	Dredged.	*	5 & 6
46	2	,, 17	25 52 0	19 22 0	1945	Globigerina ooze.	36·8	67·0	1·02692	1·02739	Dredged.	*	6
47	3	,, 18	25 45 0	20 14 0	1525	Hard ground.	37·0	65·0	...	1·02719	Dredged.		6
48	4	,, 19	25 28 0	20 22 0	2220	...	...	66·0	...	1·02720		*	6
49	5	,, 21	24 20 0	24 28 0	2740	Red clay.	37·0	68·0	1·02744	1·02753	Dredged.	*	6
50	6	,, 23	23 14 0	28 22 0	2950	Red clay.	37·0	69·2	1·02745	1·02760			6
51	7	,, 24	23 23 0	31 31 0	2750	Red clay.	36·9	68·0	1·02609	1·02763		*	6
52	8	,, 25	23 12 0	32 56 0	2700	Red clay.	37·0	67·0	1·02613	1·02773	Dredged.		6
53	9	,, 26	23 23 0	35 11 0	3150	Red clay.	36·8	69·0	1·02653	1·02778	Dredged.	*	6
54	10	,, 28	23 10 0	38 42 0	2720	Red clay.	36·5	71·0	1·02753	1·02774		*	6
55	11	March 1	22 45 0	40 37 0	2575	Globigerina ooze.	36·5	72·2	1·02621	1·02767	Dredged.	*	6
56	12	,, 3	21 57 0	43 29 0	2925	Globigerina ooze.	36·9	73·0	1·02641	1·02761	Dredged.		6
57	13	,, 4	21 38 0	44 39 0	1900	Globigerina ooze.	36·8	72·0	1·02695	1·02777	Dredged.	*	6
58	14	,, 5	21 1 0	46 29 0	1950	Globigerina ooze.	36·8	74·0	...	1·02756	Trawled.		6
59	15	,, 6	20 49 0	48 45 0	2325	Globigerina ooze.	36·2	72·5	1·02616	1·02768		*	6
60	16	,, 7	20 39 0	50 33 0	2435	Globigerina ooze.	36·2	74·0	1·02751	1·02770	Dredged.		6
61	17	,, 8	20 7 0	52 32 0	2385	Globigerina ooze.	36·5	74·0	...	1·02766		*	6
62	18	,, 10	19 41 0	55 13 0	2650	Red clay.	36·0	74·0	1·02615	1·02732	Dredged.	*	6
63	19	,, 11	19 15 0	57 47 0	3000	Red clay.	35·5	75·0	1·02614	1·02728			6
64	20	,, 12	18 56 0	59 35 0	2975	Red clay.	36·0	75·0	1·02727	1·02727	Dredged.	*	6
65	21	,, 13	18 54 0	61 28 0	3025	Red clay.	35·5	76·0	1·02688	1·02685			6
66	22	,, 14	18 40 0	62 56 0	1420	Pteropod ooze.	38·4	76·0	...	1·02698	Trawled.	*	6 & 7
67	23	,, 15	18 24 0	63 28 0	450	Pteropod ooze.	...	76·0	...	...	Dredged.		7
68	23A	,, 15	18 26 0	63 31 15	460	Pteropod ooze.	...	76·0	...	...	Dredged.		7
69	23B	,, 15	18 28 0	63 35 0	590	Pteropod ooze.	...	76·0	...	1·02693	Dredged.		7
70	24	,, 25	18 38 30	65 5 30	390	Pteropod ooze.	...	76·0	...	...	Dredged.		7
71	24A	,, 25	18 43 30	65 5 0	625	Pteropod ooze.	...	76·0	...	1·02704	Dredged.		7
72	25	,, 26	19 41 0	65 7 0	3875	Red clay.	...	76·0	1·02631	1·02692	Dredged.		6 & 7

FIGURE 7. A page from the HMS Challenger's log.

FIGURE 8. The Nansen hydrographic sampler.

Technology progressed, and as it did data types and data volumes began to increase. You're all, I think, acquainted with the Nansen bottle (Figure 8), an instrument for collecting a sample of water at any depth to the bottom of the ocean. Attached to the bottle are the thermometers that provide a record of the temperature and pressure at depth of sampling. This bottle has now progressed to an in situ, continuously recording, salinity-temperature-depth instrument that provides data in quantities never expected in the past (Figure 9). The anemometer (Figure 10), a perfectly reliable and commonplace instrument, records environmental events like tornadoes, a phenomena that has plagued our nation within the last few years. Using the lead line (Figure 11), a man made bathymetric determinations by throwing a weighted line over the side, the lead

FIGURE 9. Salinity-temperature-depth recorder.   FIGURE 10. Shipboard anemometer.

weight possibly held some tallow. He then recorded the depth and a description of the sediments. The lead line has now been replaced by the echo sounder, which provides literally millions of miles of analog records (Figure 12), non-computerized records, that—at least until recent years—were laboriously read, depth-by-depth, and plotted on charts. The advent of the telescope, the camera,

FIGURE 11. A lead line cast.

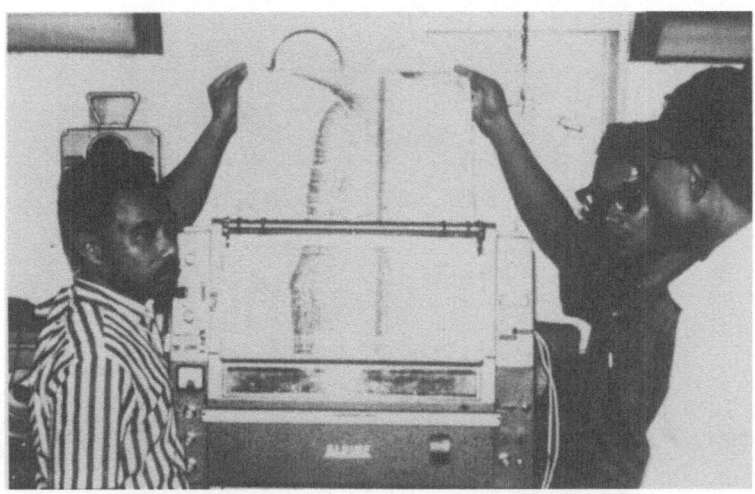

FIGURE 12. Fathometer tracing.

and infrared scanners are allowing us to take pictures of solar events (Figure 13). These are all non-computerized records.

One development that has impact on data management is the satellite. When you think that a satellite can look at all of the oceans of the world two times a day and when you think of the number of photographs and analog images that are provided by the satellites, you are coming into totally new management concept problems, a decision-making process. Which of these records must be automated? Which of these records must be saved? Think of the volumes for a decade. It's rather frightening. One development that suggests all isn't lost is that we now can use laser technology for mass storage. Each of the elements can handle a trillion bits. (Here I am, going into computers, and my topic is to stay away from computers.)

When man began his study of the environment, he was working alone as a scientist. He began working with fellow scientists; university and industrial laboratories followed. We have now moved into the era of the big experiment. A superb example is GATE—which is the GARP Atlantic Tropical Experiment

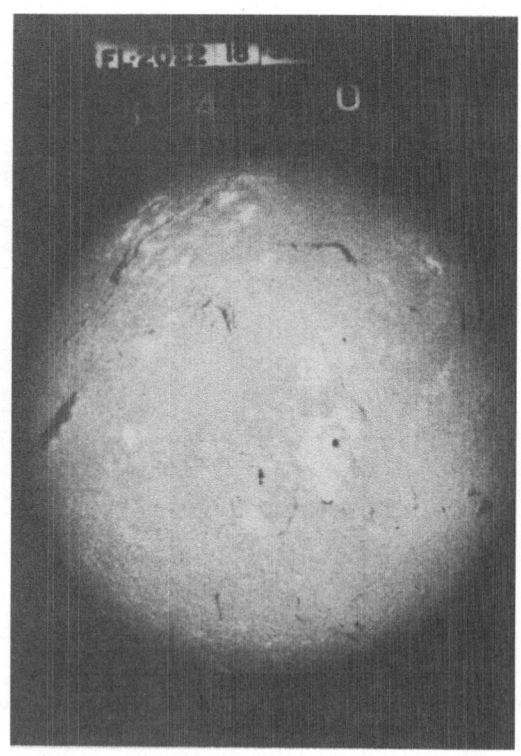

FIGURE 13. Solar flare.

FIGURE 14. GARP Atlantic Tropical Experiment.

(Figure 14). GARP is the Global Atmospheric Research Program; an acronym within an acronym. GATE involves 30 nations, 400 scientists, and approximately 3,000 people in support of these 400 scientists. Planning sessions for data management aspects of GATE have been held for the past 2 years so that at the end of this international effort—using ships, buoys, aircraft, balloons, and satellites—the data will be available as quickly as possible to the scientists. GATE is an ICSU (International Council of Scientific Unions) initiative.

Now, I will speak about some of the files within some of our Centers. By no means, of course, could we hope to have all of the data in our nation (or in any nation) within these Centers. Mr. Noel spoke this morning of a concept called ENDEX. ENDEX is a systematic, area-by-area, discipline-by-discipline (through interviews primarily) finding out of who has what environmental data. How many? What mode? Documentation? Instruments used? Calibration techniques? How do you get copies of the data? This systematic approach will allow us to bring those data into our national centers that we wish to move into our national centers. Equally, this approach will allow us to tell you as a user what data are available that are not in, and may never be in, our centers. Now we have fairly thoroughly ENDEXed (E-N-D-E-X is the acronym) the Great Lakes, the New York Bight, and Chesapeake Bay and are now moving to provide an ENDEX, or index, of what environmental data are available among the major marine laboratories in our country—Woods Hole, La Jolla, Seattle, and many others (Figure 15). Mr. Noel provided me with some statistics that I think are interesting. Out of 441 entries for the Great Lakes, Chesapeake Bay, and New York Bight, 122 are for automated data. Which means that 319 are non-computerized. Approximately 70 per cent of the data in this particular small

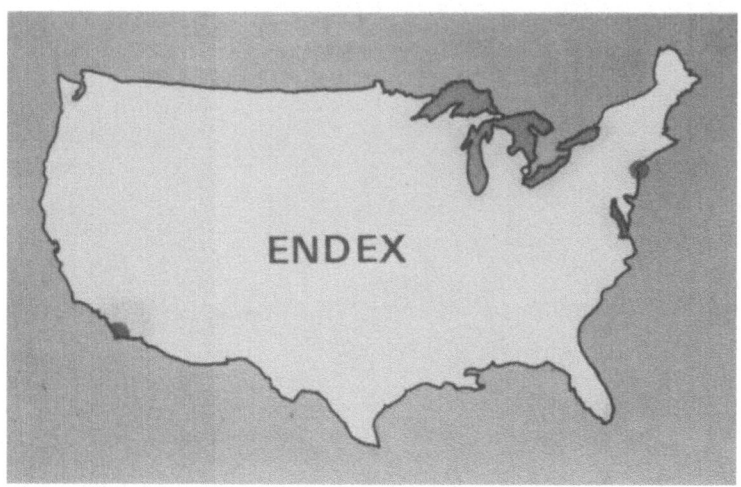

FIGURE 15. ENDEX, an environmental data inventory index.

sample of ENDEX entries are non-computerized. These include a 20-year summary of the tributaries to the Chesapeake Bay, including temperatures, salinities, sediments, and the standing crop of the biota in 25 notebooks, which equals 75 inches of log books; 33 years of air temperature, observed and recorded every 2 hours on Lake Huron on 2,000 log sheets; and 7 years of the duck mortality near Detroit, Michigan, in the river, 42,000 ducks' worth, or 252,000 measurements in a 300-page report. All are non-computerized data.

Internationally, thanks to Messrs. Peluchon (France), Kohnke (FRG), Dr. Colbrook (Sweden), and others, there has been developed an international type of marine inventory called ROSCOP, a Report of Observations/Samples Collected by Oceanographic Programs. Mr. Kohnke and his ad hoc group have been concerned with developing a second-level inventory for marine geology and geophysics. Dr. Colbrook and his group have been concerned with that type of inventory for marine biological observations.

Now, let me review with you, quickly, some of the data in our centers that are not computerized. Please do not think that these are not going to be computerized, or that a lot of them have not already been computerized. (I wish Mr. Noel and I had the time to develop with you what we have accomplished in this area and what we hope to accomplish.) For the atmospheric—the climatological data—first-order stations in our nation make observations every hour. There is also a total of 10,700 cooperative stations where surface meteorological observations are made. These move into the National Climatic Center (Asheville) in log form. These logs fill the shelves (Figure 16). You can imagine the difficulty of finding the proper log, annotating the data therefrom, and then being sure the log is refiled where it belongs.

We are moving in the direction of microfilm and microfiche. By so doing, we are providing a mechanism through what we call the hot line so these non-computerized data forms are made available, and you, any place in our nation, can call a number and request climatological data—the temperature, the probability of rainfall, or what-have-you for the location of one of the first-order stations or the 10,000-plus co-op stations. This is what happens as a result of putting these log sheets on microfilm and microfiche (Figure 17).

FIGURE 16. (Above) Meteorological data logs at the National Climatic Center.

FIGURE 17. (Left) Microfiche records.

Other nonautomated, non-computerized records in the National Climatic Center that need various types of management so that they may be made available economically and quickly are exemplified by radar and aircraft photographs. As I indicated earlier, satellite imagery is another type of nonautomated record that is kept in our centers. Also, we keep records of flood levels.

An exciting thing has happened within the past few weeks. A consortium of eight oil companies in our nation have been collecting surface meteorological and oceanographic data from six towers stretching across the Gulf of Mexico (Figure 18). We have been negotiating for these data and will have them available shortly. One of the interesting things, among others, is that a hurricane passed over one of the towers; the data were recorded continuously, once every second. In at least two other cases, hurricanes passed in between the towers.

FIGURE 18. Offshore oil platforms collect meteorological and oceanographic data in the Gulf of Mexico.

Next, I will speak about the oceans. NODC's BT's and Nansen bottle station data were received from national and international sources through cooperative exchange. Through 1969 there were 214,000 BT's in analog form (Figure 19). They're no longer in analog form—they have been digitized. Nansen casts, which, of course, is the classical method I mentioned earlier, numbered 124,000 between sea surface and the bottom or intermediate depths in 1969. These were provided in analog form, but have now been digitized. The new in situ salinity-temperature-depth recorder is lowered from the ship like a yo-yo. In going

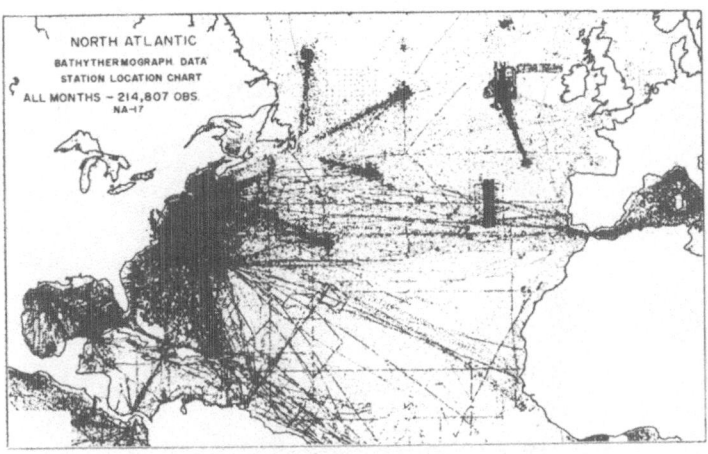

FIGURE 19. Archived BT Observations, North Atlantic
Locations (NODC - 1969).

up and down, an analog (Figure 20) is recorded on shipboard. This opens up a whole new problem. A scientist at Woods Hole was interested in some of the records taken with salinity-temperature-depth recorders. He was interested in data for 5-centimeter intervals between the surface of the sea and 3,000 meters, a request that required extensive digitizing.

Now, one quick deviation from non-computerized data to results of automation. To use that data that I mentioned for the BT's, Nansen casts, and a number of other types, we have had to go to computer graphics (Figure 21). Which means, of necessity, the data are automated. Once the data are computerized, the oceanographer or the scientist who has access to the computer can interact directly with the computer in what we are calling the "live atlas" concept (Figure 22). The user may call the data forth: whatever depth, whatever area, and whatever

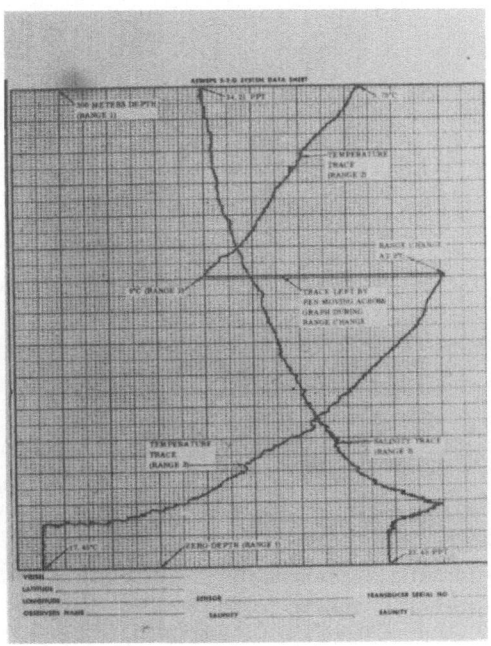

FIGURE 20. A shipboard analog of salinity-
temperature-depth recordings.

FIGURE 21. Computer graphics –computer output of data in
graphic form.

FIGURE 22. NODC's Live Atlas—televised output direct
from the computer.

parameter is in the files. Once the   data are on the CRT, they are in a non-automated analog form.

Let me now speak about the satellite problems I mentioned earlier. Figure 23 is an HRIR picture of the Gulf Stream—the East Coast of the United States is clearly shown, the Labrador Current is coming down the coast, the colder water is

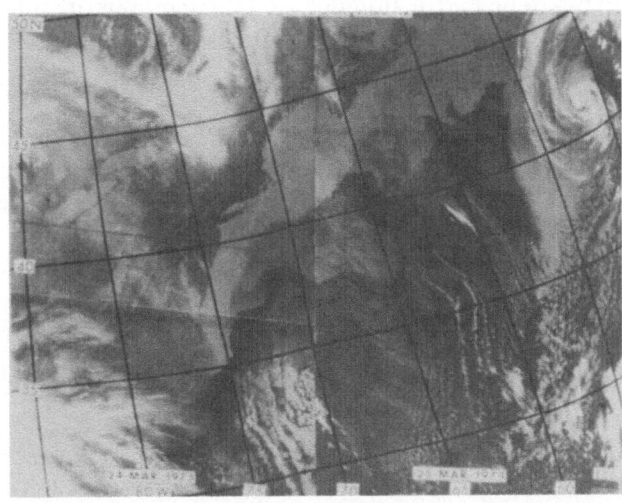

FIGURE 23. A high resolution infrared (HRIR) satellite
picture of the Gulf Stream.

the light area, and the warm water of the Gulf Stream is the dark area. Satellites provide thousands and thousands of these images. Decisions have to be made as to storing versus purging. At the moment, there is an international exchange agreement involving satellite imagery of waters off the West Coast of Africa. Participants in the International Decade of Ocean Exploration, along with members of the Intergovernmental Oceanographic Commission and the World Meteorological Organization, are conducting surveys of the upwelling off Dakar. We are providing satellite imagery, which will be compared with the ship-collected ground truth data.

I mentioned files for bathymetry, geology and geophysics. These are files for which automation techniques are less applicable. For example, there are millions of miles of bathograms, echo sounder records (Figure 24). The echo sounder record is read, then written on the boat sheet along with latitude and longitude determined with a high degree of navigational precision (Figure 25). These are then brought home and used to prepare bathymetric charts (Figure 26). The boat sheets are kept. They have become so voluminous that NOAA has now developed a technique for digitizing (Figure 27) the data from the boat sheets, which are then converted to microfilm and microfiche to reduce storage space requirements.

To acquire solid earth data, a world-wide cooperative network of stations record during quiet seismic times as well as during any major events like earthquakes and tsunamis. These provide endless yards of analog records and seismic chips (Figure 28). During the earthquakes on our West Coast not too long ago, we very fortunately had a number of seismographs in operation along with accelerographs. One of the disquieting facts is that the accelerograph near the epicenter recorded 6G's.

There are now about 21,000,000 nautical miles of shipboard underway records of geophysical phenomena from the oceans of the world. The seismic reflection recorder is towed and records continuously while the ship is at sea.

FIGURE 24.   A bathogram record.

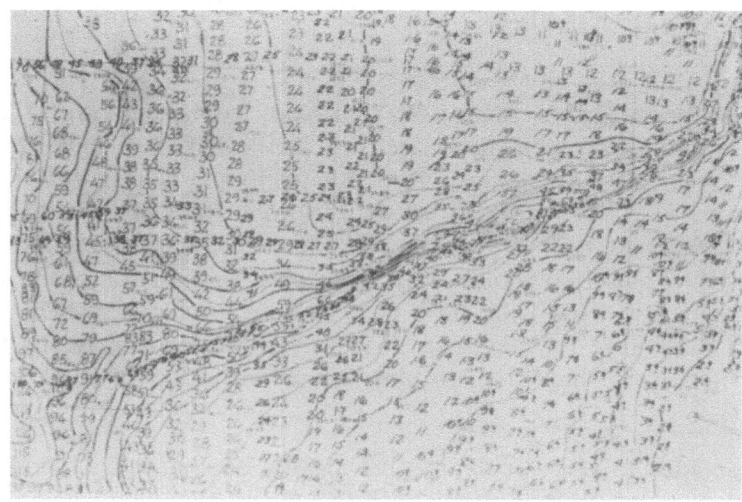

FIGURE 25.   Worksheet of bottom contours.

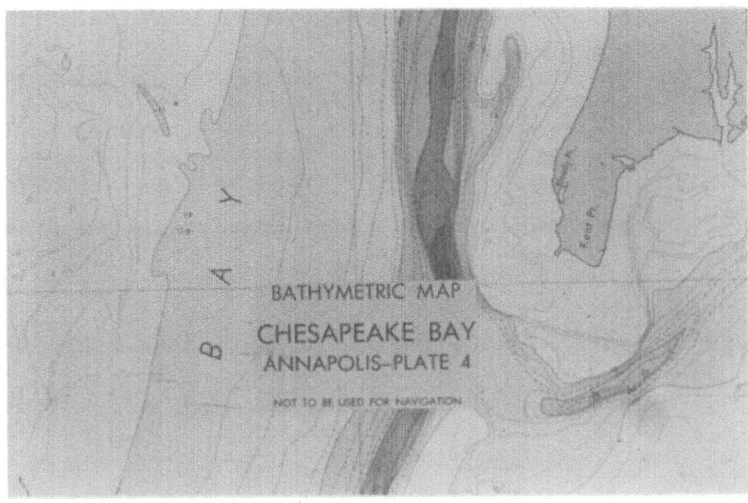

FIGURE 26. Bathymetric map of part of Chesapeake Bay.

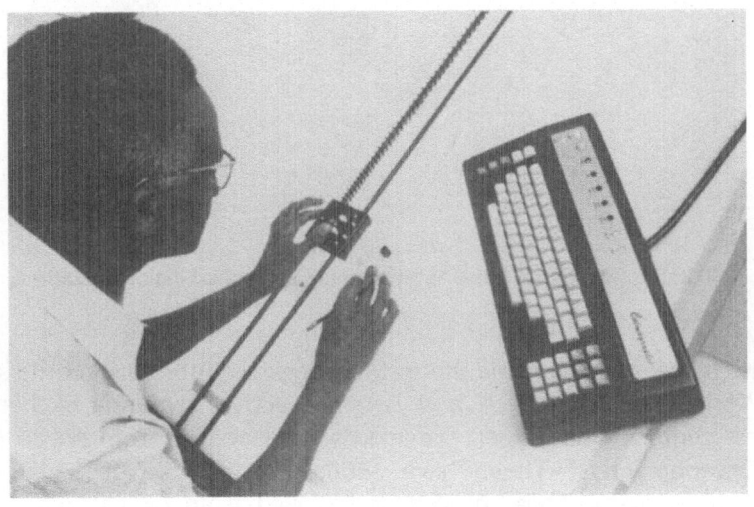

FIGURE 27. Digitizing bathymetric data.

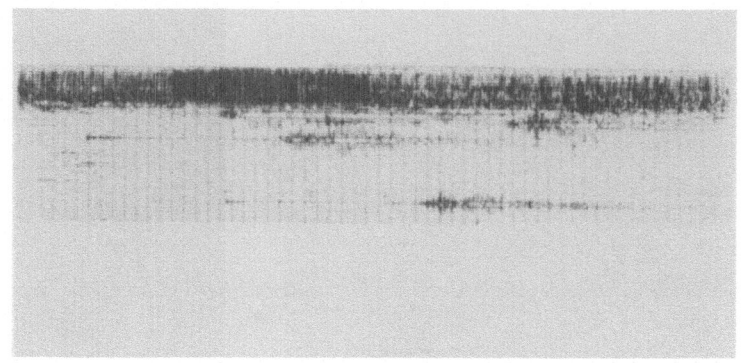

FIGURE 28.   A seismograph record.

FIGURE 29.   Cooperative international data collecting organizations.

The name of the game here this week is international (Figure 29). I mention World Data Centers A and B, since active, dynamic exchange goes on among these centers. I have been provided some statistics from World Data Center A for Oceanography. They have 800,000 marine observations, all in non-computerized form. These have been received from 55 nations, which means 55 of the 76 member nations of the Intergovernmental Oceanographic Commission have provided oceanographic data—marine observations, if you will—for international exchange. Eighty to ninety per cent are nonautomated, as of now, within World Data Centers. All environmental disciplines are involved: atmospheric, terrestrial, ionospheric, solar, glacial, geodetic, and tsunamis—each with its own World Data Center—brought together under one management.

International exchange is valuable for a multiplicity of purposes: knowledge, scientists speaking to scientists, and scientists making their data available to other scientists. You can argue that any scientist knows the rest of the scientists in his field and can write to John, or Bill, or Joe, or Dieter, or whoever they happen to be. True, as long as you're alive; but we're all going to the happy hunting ground of data managers one of these days, and unless our data are completely documented and can be made available through appropriate centers, they're going to be lost for the coming generations of scientists. The International Council of Scientific Unions, the International Union for Geodesy and Geophysics, and the International Association for the Physical Sciences and Oceanography are groups of scientists who, without exception, are involved to some degree in formal data exchange, and I don't mean scientist-to-scientist, I mean through formal data exchange mechanisms. It's becoming vital that we make these data available to the scientists in the developing countries. To a major degree they are made available in non-computerized forms, but that's beginning to change and change rapidly. To an increasing degree, the developing countries, as far as environmental data management proclivities are concerned, are becoming increasingly automated and sophisticated. However, we still have to be able to provide the data to the user in the form that he or she needs. (You notice how I slipped in that "he or she." Women's Lib is getting to me.)

Also, we need to provide environmental data in terms of services. By service functions, I mean such as global weather; United Nations Environmental Program, discussed this morning by Dr. Peachey from GEMS; and the Global Environmental System. These data will be of no use unless there has been, from the beginning, a data management plan developed in the international/ intergovernmental arena . . . otherwise, the data either will remain with the scientists who were involved or they will be lost; there will be no proper documentation. We are now moving into an era that necessitates international/intergovernmental data management and exchange. The data are becoming available in totally hitherto-unknown quantities, for instance, from GATE. The GOES satellite data alone will be on 3,000 magnetic tapes of 2,400 feet. And that's only one type of observation among the satellites, ships, aircraft, buoys, balloons, and what-have-you. We are becoming global minded and less provincial and nationalistic when we think about our environment.

I'm very pleased to have been involved with this group, and I thank you.

DISCUSSION

CUTBILL—Just for a moment I am going to put on my geology hat and revert to one of those types who wander around the countryside with a hammer, a notebook, a hand lens and a scientific tape measure and blow dust off of fossils in museums. I am horrified, I really am. I haven't realized the scale of all that is going on, and I have just got a sneaking question: which came first, the computer gadgetry or science? Are we taking advantage of a tremendous ability to collect data and just doing the kind of things that that releases to us, or is there somewhere in the organization one that is spending all of this money for some planning, for some thinking about what science (what things like this) are really good for us in this field? Would you have a crack at that for us?

AUSTIN—You used the word "sneaking" advisedly didn't you? Well, I like to think that the science developed like the man pictured in the dunes. He learned more and more about the sand area through which he moved, or whether he happened to be taking a hammer to the mountains and knocking off pieces of rocks, whether he happened to be involved with weather, or whatever he was interested in. I like to think that he began to work with others, either in a university with colleagues or with a funding agency, to get support for his kind of work as it became more and more expensive and because he wanted to move about further and further.

What I am saying is that I think research has evolved because of global interest in the weather, because of interest in the North Atlantic Ocean and its distribution of fishes, or just because of human curiosity. People like the Joe Reeves, and other people in Oceanography, could bring the data together and use it to describe phenomena in the North Atlantic or Pacific. Man is beginning to wonder how fast the currents move, what are the tidal relationships, what is the relationship between the Trade Winds and the return flow in the Equatorial Countercurrent or the Equatorial Undercurrent, what is the relationship between cloud clusters, latent heat, the mixing layer in the development and existence of a hurricane, and so on. These are things that no man can solve alone. He no longer is an island unto himself. He has to begin working with his fellow scientists.

So it has been a slow growth. How slow is in the eyes of the beholder. If it is true that there are more scientists alive now than the total that lived during previous recorded history, then perhaps the growth is asymptotic; I'm not sure. If you, being inquisitive, want to compare what is happening in the Rockies with the pieces of stone that you have broken, you can, if you have properly documented your observations, see what the relationship is, if any, with orogenic events in the Himalayas or the Alps, as you begin to broaden your interests.

CUTBILL—You have done a wonderful description of scientific curiosity and where it leads, but should we be making other kinds of decisions such as whether that is the right kind of scientific curiosity?

AUSTIN—I can't answer that and I would rather not even try because I think that is in the eyes of the beholder. If you have a keen interest, as we all should have, in what is happening to the environment and, let's say, through man's inadvertent effects on the environment, you are going to have a totally different type of approach. You are going to be keenly interested in UNEP and GEMS, you will want to see what is happening to our environment, what the trends are, and how we can reverse them.

Our friends here in EPA are deeply concerned with the regulatory aspects in our nation, but they have to know the trends, they have to have a baseline. They either have to establish it because there are inadequate data or they come to the source data and say let us see what the baseline is. They then begin monitoring, and when the trend reaches a certain level, the red lights flash and whistles blow. We have got to reverse that trend.

If you are interested in what is happening to resources globally, that is a totally different type thing and calls for a man with a totally different type interest than the geologist with his hammer or the person monitoring biological oxygen demand because of a pollution problem. We are keenly interested in how are we going to more reliably predict tornadoes. Look what happened in our nation in the spring of 1974. We were becoming rather smug about the way in which we had been able to predict tornadoes and minimize the losses. Well, that smugness is gone. People for whom I work and with whom I work are out in airplanes and helicopters talking with people and trying to find out what we did wrong, if we did anything wrong, or how we can improve. So to each his own.

OPPENHEIMER—What about the possibility of reducing environmental data to a weather map concept? I get the feeling from your talk that one of the major needs that we have today is the compressibility of data. So that the digit information point may be expanded to a three-dimensional system where you use a three-dimensional laser to identify points in which you can then compress your information into very small units.

AUSTIN—It might not be a good answer to your question, but we publish what we call the LCD's—Local Climatic Data. These are updated every ten years for any one of the first order stations that I showed you on that chart, plus a large number of cooperating stations. It's done by a technique called PERSUM, which is PERiodics SUMerization of climate. This provides for a particular station the average, the maximum, and the minimum temperatures, the rainfall, the snowfall, the cloud cover, etc. This is a way to compress and make the data readily available to a wide number of users.

OPPENHEIMER—I was thinking more of temperature, salinity, mercury, etc., in the oceans, in a three-dimensional sense in the computer for analysis.

AUSTIN—We can do that to an increasing degree with computer graphics.

I think more and more, Carl, that user demands are so different that I am not sure atlases are cost effective any more. Dr. Colebrook and others have heard me speak about this, apparently ad nauseam. But when I think it costs $100,000 to publish the Indian Ocean Atlas, I have to be sick. All of those data are on disk and mag tape and available to use in any way you wish. The atlas looks awfully nice on your coffee table, but I am not sure how often you would refer to it. And I feel that, well, holography is certainly coming into its own, and eventually laser holography will show three-dimensional distribution of properties. If that is what you want, that is what we can provide; but you had better have a pretty good contract to support the cost of retrieval.

PEACHEY—Mr. Chairman, I think we have again touched on the problem of what it is reasonable to monitor to protect a nation and a world, plus what it is scientifically desirable to measure, plus what it is curious to measure. The kind of organization which springs up or develops to encompass all these things is going to be clearly different from that which, say, may develop where concern is with very limited goals.

I think a part of the proceedings of this Conference might highlight the philosophical differences which affect people when they are either in this general service function or in the straight jacket of a regulatory or an executive management system. Of course, we need a note of caution about all being obsessed with good management. Human beings do borrow sideways from experience gleaned speculatively, curiously or otherwise. I was just trying to think of cases where a very high concentration of resources on the maintenance of historical collections has proved profitable.

For a time, I worked at Rothamsted Experimental Station in England, which is known for its carefully maintained plots on which certain soil treatments have been maintained constantly since the last century, and, naturally, there was a continuing debate on how sacred these plots were. But when you need to resort to either living or dead biological museums for background levels of pollutants and other parameters, the issue is plain. Always, we shall be faced with how much energy to put into the archive and how much study to do for study's sake. It is a basic fundamental, intellectual problem. The relevance to this present meeting is presumably: does it affect the different arrangements and outlets one has in order to manage the system, whether it is an intensively mission-orientated system or whether it is archival? Would you say that you have described basically a library system, in the best sense of the word?

AUSTIN—Well, having lived with a librarian for 30 years, I'm not quite sure how to answer that. If the library's primary purpose is to serve, whether it happens to be the public, or scientists, or industry, yes. But a very definite decision has to be made.

The Department of Commerce is one of the Executive Department levels in our Government, as you know, as are the Departments of Transportation, Health, Education and Welfare, and others. Within the Department of Commerce, one of the operating agencies is NOAA, the National Oceanic and Atmospheric Administration. When NOAA was formed in 1970, the Administrator, Dr. White, made a conscious decision. The major line components of NOAA are National Environmental Satellite Service, National Weather Service, National Ocean Survey, National Marine Fisheries Service, Environmental Research Laboratories, and Environmental Data Service. A series of questions had to be answered—Should there be a mission-oriented environmental data center within each of the major line components? Should there then be a czar on the staff who would keep an eye on each of these centers to assure a systems approach? Or should the centers be brought together in an environmental data service? Dr. White decided on the latter. So we have the five centers—the National Climatic Center, the National Oceanographic Data Center, the National Geophysical and Solar-Terrestrial Data Center, the Environmental Science Information Center, and the Center for Experimental Design and Data Analyses. A new one is being formed, the Center for Climatic and Environmental Assessment. This is to be a group of very specialized scientists who will look at both the modeling and the continuous verification and evaluation of climate interaction with crops, energy, fishes, etc.

The Environmental Protection Agency, the Department of Defense through its Office of Naval Research, the National Science Foundation, the Department of the Interior, the Geological Survey, the Bureau of Land Management, universities, and industry have very formal, bilateral agreements with the Environmental Data Service. The agreements are signed at a reasonably high administrative level so that they are gospel and say that, first, our Data Centers will be provided, within three months, with a description of any activity that results in environmental data. And these data will move into our Centers within one year and that the expense of preparing these data to move into our Centers are the agencies' expense, not ours. Once they are in our Centers, then it is our responsibility (and at our expense) to keep them in a form so you as a user can get them out of the Center. Our budget in base is about $12,000,000 a year; we have about 660 people. So NOAA's administrator has put the resources there. He is providing the resources for us to do the job that we feel has to be done and that he feels has to be done.

You were asking earlier about the structure of this. It has to be a national policy, I think, or it isn't going to work, and to a degree it is, if you will, a national policy. So you as a user pay for the cost of retrieval only; you don't have to pay for the cost of getting the data into the centers or maintaining it there. You pay only the cost of retrieval.

OPPENHEIMER—I would like to follow up what Dr. Austin mentioned earlier about the cost of going into three-dimensional retrieval. Lack of environmental facts, it seems to me as I work with environmental data and groups who need the environmental data for their environmental impact statements, may be responsible, in part, for the big energy and monetary crunch that we have. And I use as an example the Northeast Gulf of Mexico Offshore Lease Agreement that was assigned just a few months ago, which cost the oil companies, if I remember my figures right, about $800,000,000, and it seems that they are not going to be able to exploit their leases for almost a year because of the Bureau of Land Management's restrictions based on insufficient data. The interest alone on this investment is $90,000,000 for just that one lease area. If one multiplies the loss of time, the use of money or resources for our coastal environment or our fisheries around the United States and all over the world and puts it in a monetary loss mode, then we do have a very good selling point to say that the small cost of three-dimensional data retrieval can save time, energy and funds, and may save our environment. It is of consequence.

AUSTIN—To a major degree, three-dimensional retrieval is gilding the lily. You can see what you need without it.

OPPENHEIMER—I was only using that as an example. Environmental data systems do not have to take second place to any current science field because they are absolutely necessary.

AUSTIN—Somebody mentioned earlier this morning, and I will express it bluntly, that data management ain't sexy. I have heard that a lot in the last two or three years and it is true. If there are twelve items that are ranked according to priority of expenditure of resources, more often than not the data management aspects are number 12. Because if we don't do something this year, it isn't going to make much of a difference if we wait and do it next year, if resources are tight. So data management is a full-time job of salesmanship. A group like this is, I think, superb because, hopefully, what will come out of this is another item that I can wave and say, look, here is a gang of experts and they agree.

OPPENHEIMER—But one must come down to reality. For example, it is a lack of communication of information that is holding up the development of rivers or dams that are going to be needed for water for an increasing population. Another example is the development of energy. Construction of nuclear energy plants is being held up because of environmental impact statements.

AUSTIN—Let's not get too carried away because a lot of that holdup is eco-emotion. You know it as well as I do that more injunctions are holding up what you are talking about than the lack of environmental information. So it is a problem, and we have got to face it, but eco-hysteria enters into here also.

BROGDEN—I would like to start off in another direction here because something I was hoping to get out of this afternoon's session is a better idea of the role of institutions, the older institutions such as museums, which may or may not be computerized, or are starting to computerize. They have tremendous non-computerized information and data bases and I just want to know more about (particularly since we have a museum person here) what role the museums see for themselves in this sort of thing and who they consider their users to be.

CUTBILL—Well, as I am a museum person and as there aren't any other museum people here, I don't think, I can afford to be extremely rude about them. I don't think they know what their function is. I don't think they know what public they are serving and what the evolutionary framework is. As a result, of course, they can't make good cases for money and they don't get money.

To my way of thinking, they have very obvious and important functions because they have from the environmental point of view an overall responsibility for preserving actual, physical objects from the environment for the future. The rationing that now goes on of pre-1945 animals is a case in point. The samples which were made in the past were totally inadequate. And you get the museums with shortages of funds; they know their records are bad, they know that they are not properly managed, but they don't have the funds to do much themselves and they don't make the arguments for them. It's changing slowly. I think there is now much better realization of this problem among the museum community, but ultimately the people who will make the difference to this are the museum users. It's having customers who are requesting answers that really puts the people who are doing the management on the spot.

SCOTT—I want to say more about the role that we see the museums in. We are trying to set up a network of local environmental records centers in England and we see the museums as fitting into this pattern. If anybody is interested we have a publication (Centers for Environmental Records, 1973) which contains the proceedings of a conference we held last year for various people throughout Britain who were trying to set up local records centers. We have succeeded in getting all the people interested, but as John (Cutbill) said, the problem is in funding. We would like to fund them, from our end if possible, to help them. But they definitely do have a role.

AUSTIN—I think it is a new concept and I would say, as long as there are no leading museum people here, that your approach is novel, that museums lack the type of leadership we are talking about. I am not sure to what degree a man who has spent his whole life in the Himalayas counting the feathers around the eyeball of one bird and who is now the director of a museum is going to be interested in bringing this type of data to the public. My daughter works in the Smithsonian in Washington and spends long hours sitting there painfully writing numbers on the bones of a skeleton of a bird about the size of a humming bird. Well, whether or not that can be done more efficiently, I am not sure, but that is the way it was always done and that is the way it is going to be done until young people like yourself and Miss Scott and others become involved.

ROSENFELD—I suppose that while we are pursuing this discussion we ought to ask whether we are talking about vast differences in kinds of data between a museum and a satellite. In fact, I know what the answer is, and it will do us no service to be discussing those in the same context. You are talking about a geologist going out and picking up his rock or the biologist numbering the bones of his bird at the same time we are trying to discuss the satellite taking millions of bits of data. Somehow or another I think they are different in kind, and really it's not a continuum just based on data quantity. We ought to focus at some time during this Conference on which kind of data we are talking about or whether we are talking about both kinds of data and how we are going to handle them.

CUTBILL—I think this brings out a very important point as to whether these are truly different kinds. As a museum man I would perhaps doubt this. The quantity of data is certainly different; we can't compete with the satellites. There are something like 500,000,000 museum items in Great Britain and the growth rate is on the order of 2,000,000 items a year. Now, they are not small numbers on any scale and the kind of information covers a very wide range from quite numerical scientific observations to totally unscientific sets of data. There are differences of scale and differences of organization, and it is not so concentrated, but I wonder whether it's really that different, whether there are different management problems, whether there are different positions.

PEACHEY—I think it comes back to this fundamental point of whether or not you are performing a general service function. A satellite is basically there for some other purpose, but it's also collecting everything else as well in a kind of new Victorian collecting spree. The general move towards accountability means that whenever we come across something like a data management project that doesn't have a very good picture to display, we always dress it up as a specific management tool, and we never expose it as a general service, even though it may be one. In this way, we manage to get money for it. People basically don't want general service functions. They just want to feel better.

CUTBILL—What are you suggesting we should do about this?

PEACHEY—To identify where we must have a broad and common data base for management and in support of policy, they will inevitably contain a general service function. The real problem is to what extent are we talking about the general educational obligation we have to our children and to what extent can we sell these things directly in support of what people want—bluer skys, nicer homes, pleasant back gardens, and estuaries you can swim in.

In Britain we are beginning to tie these things very close together. We have a number of Regional Water Authorities that have total responsibility for the whole hydrological cycle within their area. These new institutional arrangements should justify and produce a comprehensive data management system for each authority which is susceptible to national co-ordination and to the needs for a general service. It seems to me that it is very much easier in these circumstances to upgrade the standing of data management.

OPPENHEIMER—We specifically invited the Texas Water Development Board because they are doing precisely the same thing. Perhaps they might want to say something about this.

NELSON—I will attempt to respond along the lines of the institutional responsibilities starting with the local governments, River Basin Authorities, and continuing up the bureaucracy to State and Federal levels of water resources responsibility. In Texas, the river authorities are generally the local sponsor for water resource projects. But the tremendous costs involved in data collection programs for planning, monitoring, reservoir construction, operations and maintenance of waste and water treatment facilities make it necessary for those who benefit to share these costs.

Here, we are talking about the information required for the decision-making processes, for planning projects, monitoring environments, and evaluating the impacts of our society. Information on meteorological, hydrological, chemical and biological parameters are required at all levels of government for decision-making and environmental assessments.

With the advantage of an environmental data bank, the whole water resources economic picture for a river basin can be analyzed for municipal, industrial, and agricultural water demands and the resultant impact of return flows on the environment.

With such complex problems to solve, models can be developed using the environmental data bank for excitation and calibration of these analytical tools. The analytical tools can then be used to evaluate various management plans and criteria for working within the assimilatory capacity of the environment we are dealing with, whether it is a stream, reservoir, or estuarine environment.

So the questions are where to store and what to store to get the best reliable information in our respective data banks. This, I think, is what we are talking about today.

Who should be the responsible entity for control of the banks of information? Who is responsible for retrieval and analysis of problems; and who is responsible for selecting the analytical techniques to get the most reliable answer to our problems?

The single most important issue may be the competence of the data that is stored. Because if we don't have confidence in the data that is stored, how can we, with a clear conscience, merge these data basis' hydrological, meteorological, chemical, and biological parameters—for problem solving? The users, and this is one of the things I have seen here today, are concerned about the other man's data. How was it taken? Was he as careful as I would have been in the collection and analysis of the data? This is a problem we have in the State. And, as we all know, there is usually more than one method for environmental measurements. For example, the measurement of dissolved oxygen can be performed by a wet chemistry method or with a dissolved oxygen meter. Now, the question is the difference in the two numbers we obtain. Well, in the realm of environmental use criteria for determining the environmental limits of the organisms that we are responsible for protecting in this State and the Nation, the limits are generally wide enough so that a difference of a half part of oxygen isn't going to destroy the population.

So, we must take into consideration the concentration at which the parameter is toxic to our area of interest; in our instance, Texas, and I am speaking of general environmental issues, not the controversy of what is toxic and what is not. We have to determine what these environmental limits are and what effects they have on a stream or any part of the ecosystem. In the evaluation of an ecosystem, we must have a systematic approach to define the problem of a river basin and the local entities within that basin.

Then we have the problem of overlap in governmental responsibilities because the environment doesn't know about state and county boundaries, and in Europe the water and organisms don't limit themselves to the different countries. It would be nice, of course, if a mountain range divided each governmental boundary so that the hydrological systems would be unique for each governmental entity. But it doesn't happen that way. Generally, a river system crosses many governmental boundaries as it works its way through the land to the coast. So co-operation between governmental entities to solve our complex water resources problems is a necessity.

I don't know if I am off on a tangent here, but these are some of the basic issues I see in handling an environmental data information system. I hope to demonstrate later a case where we developed a storage and retrieval system to satisfy some of these questions. We use this system in developing our environmental assessments of water resources development projects to determine their impact. I could go into all the models we are developing, but it is the parallel analytical tools that I think are the most important. We have a basic environmental mathematical model we are developing, but this model is actually many small component models linked in an interactive manner.

In the development of our parallel models, we are exciting these models from an environmental data bank of information. These data are also used to calibrate our model simulations. We break the problem down by looking at one parameter, playing it against other parameters and seeing how it responds. For example, how does an organism or group of organisms respond to environmental influences? With the ability to lump organisms into similar groups—we talked about the lumpers and the splitters this morning—we have to lump things into logical groups to be able to handle a complex environment in an ecological model. We are modeling two phytoplankton types, the cold-water organisms, diatom-like organisms, and warm-water organisms, blue-green and green-like organisms. But their environmental limits are such that we can reduce the environment to an easy-to-handle base.

Therefore, we can develop our technology and our simulation capabilities so we can start asking ourselves questions. What if you build a dam here as opposed to here, or build a nuclear plant, what will be the environmental impact on the estuary, stream, or reservoir?

AUSTIN—Are these Regional Water Authorities an actuality or a concept?

PEACHEY—The Regional Water Authorities are set up. But I think what we have just uncovered, and I think the Texas experience parallels this, the need to study the way in which institutional arrangements affect data management. Because it would seem to me that it appears to be a much more dominant factor than one imagined. Lower down we come to unifying concepts. Obviously, if you are allowed, whether in a local situation or a regional one, to look at a whole ecosystem, then that is going to dominate the way in which you organized your whole data collection, measurement and management. I wonder whether the Battelle thinking has been to take things as a whole in this way. Could you say if I am right?

KOVACS—Do you mean as far as an environmental impact statement? While I am not in the Environmental Sciences Section of Battelle, and instead deal with information systems primarily, I have observed that Battelle views the problem in its totality. Actually, under four labels: ecological or physical factors, aesthetic factors, sociological or demographic factors, and economic factors. The whole problem is addressed in order to determine all impacts that will result, at least all those to a specific situation; i.e., a geographic region.

PEACHEY—And it would be interesting to compare data management within an organization that has total management responsibilities for identifiable ecosystems with a more conventially organized vertical bureaucracy.

NELSON—Dr. Peachey, your concept happens in the United States as a policy because any regional development agency for water resources projects is required to assimilate all the environmental information in order to construct a reservoir project, to building anything that might have an environmental impact. So the responsibility of evaluating the environment is placed on the developer. And I think that is where it probably should be because the developer or planner is liable to several agencies. In our case, there are six different agencies we must take into consideration for environmental assessments. We try to address all environmental needs in light of local, state and national goals. So, we start at the local level and work up through the bureaucracy regime and then back down to the local level.

CUTBILL—There seems to be a problem arising. We are discussing very recent governmental activity and by and large there is quite a close link between the organization that set them up and the purpose for which they were set up. At the same time, we seem to be certain that the environmental data bases that are being created and discussed here are going to outlive governmental restraints. Political and organizational structures are very temporal things, and there is a question as to what is going to happen to these when the management for them must be divorced from the immediate organization and purpose for setting them up. Perhaps, we should be looking back at some of the past activities of this kind where this has happened. I was wondering about Vic Loudon, who, of course, is in an organization, a geological survey, where transition has been common for a long time. Perhaps we should consider other people with a similar sort of transitional experience who might explain what happens in the long run.

BROGDEN—This is, I think, what we have in mind in our program in the rather cryptic line, "historical events for environmental systems analyses." That rather cryptic phrase was intended to start people thinking about the historical aspects of data systems.

LOUDON—I think this is a very interesting point; the question of responsibility for data collection, responsibility for data storage, responsibility for deciding what form the data take, what form the data are collected in. Our Institute of Geological Sciences, having a long history, may follow some procedures simply because "this is the way it has always been done." The unkind mutter, "sacred cows;" the benevolent, "preservation of the consistency of the data base."

To some extent we are now moving into a situation where requirements are indicated by outside ogranizations who require information of a particular kind which would normally be the responsibility of the Geological Survey to collect. A contract is then negotiated between the government department which requires the information and the Institute of Geological Sciences which has the expertise for collection of this information. To some extent perhaps this overcomes the problem which Dr. Cutbill has just raised, because the organization collecting the information can maintain consistency of data collection over a longer period than could the various "customers" of the information.

This is getting away from the questions that you raised originally, Mr. Chairman, about non-computerized data—what guidelines we could set up for masses of data and how these could be mobilized; whether it was in fact necessary to computerize data in order to make them widely available.

This is particularly significant in a situation where you have data collected over a long period of time, because a very considerable amount of intellectual effort is then needed in order to take all the old data and put them in a form which can. be read by the computer. And it is doubtful, in many cases, whether this intellectual effort is worth-while. The data may be in a manuscript form and may have been there for a hundred years with no one using them. Perhaps they have not been used because the requirement did not arise. Perhaps the requirement never will arise, or perhaps in a hundred year's time this information will become extremely valuable. So it is important to hold that original record. But it is questionable whether it is worth putting the intellectual effort into getting it into a computer form.

This raises again the question of why one wants to computerize data, what advantages accrue from computerizing them, and under what circumstances might it be acceptable to leave them in the original manuscript form.

We can see technological developments other than the computer which could mobilize this information and make it more widely accessible. Two obvious areas are microform techniques for more compact storage, cheap reproduction and communication in the sense of being able to make the data available to many users; and facsimile transmission, perhaps an area which hasn't had much impact yet, but which would enable us to communicate much more rapidly, information which never had been computerized. Another significant area may be the advent of computer indexes to data—the ability to produce, on the computer, printed indexes which people can use on their own desks. The computer does not make the search, but is used to generate a manual access tool.

There are areas where we can see that the advantages of computerization of data are relatively small. Graphic data, photographs, or maps, for instance, are expensive to computerize. The advantages of putting data into a computer-readable form must be weighed against the considerable cost of time, manpower and equipment to do this.

One of the questions that does arise in this general area of computerization of data is whether it is desirable to have data bases directly reflecting the aims and objectives of the organizations that establish them, or whether it is better to have a separate organization with the general service functions of holding data that could be accessible to many different kinds of users, and to different agencies that are using the data for their own purposes.

CUTBILL—What you have mentioned is something else that perhaps should be discussed at length and that is what appears to be a dogma around this table that we should keep data. It costs money to keep data. The more fancy gadgets we get and the more data we get, the higher proportion of our effort goes into maintenance. Should we have some sort of policy of selective data preservation, and if so, what? When do we reach a point where we have to do something about this?

OPPENHEIMER—Dr. Cutbill, your session comments are a point for discussion on the last day—that we definitely should originate a concept as to which data should be computerized and in which type of mode.

I would say that one could look at data in two modes. One is the historical, or maybe conservative, type of data like geological rock specimens. After all, a mountain very rarely changes overnight, as compared to the second type of data which is more transitional data such as you would get in a body of water in which we have seasonal and rapid daily or diurnal changes. I am not saying that mountains don't change in geologic time; I am trying to compare the two types of data—the conservative and the nonconservative.

To me, as an ecologist, there is no such thing as replication of environmental data points because of the impossibility of taking duplicate samples of the same water or air or organism. As Diana Scott said earlier today, the more data available, the better the validity. At least this has been my experience with items such as temperature, chemicals, etc. And I still like to look back at the possible concept of a world data chemical map which may compress the data, like temperature or barometric pressure in the weather system. As the data is placed in concentrated units or concentrated areas, the system can be used for those types of nonconservative information to show change. Whereas, for data that do not change, such as sediment types in an area of slow or nondeposition, we can use the historical approach or the museum approach.

AUSTIN—It seems to me that environmental impact studies have really come into their own in the 1970's. There wasn't much interest in the 60's, the 50's, or the 40's.

We are entering a whole new era during which we are not going to be able to use resources with the utter abandon of even the 60's. So, here again, we need environmental data to establish certain kinds of baselines.

We celebrated the hundredth anniversary of the AG/MET Bulletin about six months ago. Man has looked into climate, agriculture, and the weather, insofar as a formal periodical is concerned, for a hundred years. We are learning to forecast by hindcasting. We watch things happen in nature and we say if B follows A and C follows B, then there is a very good probability we are going to be impacted with D. Without the data to help us to hindcast, we cannot improve our method of forecasting, whether it is forecasting a hurricane, a tornado, a drought, or a flood.

Just plain interest in terms of knowledge, of course, has been with man ever since we can remember. We are now looking at global weather. We realize that things happening over the North Pacific are linked directly with things happening over the North Atlantic, and if we know what is happening over the North Pacific we can predict what will happen over the North Atlantic. (I once heard a paper in Dublin. A gentleman from the U.K. said that he could forecast the weather over the Northeast Atlantic more precisely on a 30-day basis than he could on a 24-hour basis because of the global linkages in the Northern Hemisphere.) In transportation, look what is happening with the need for new kinds of weather information for aviation or shipping.

All I am doing here is listing what you and I know we need right now in terms of environmental data. Who knows what we are going to need in the 80's; our needs may be totally different. There has to be a degree of calculated risk in husbanding the data. We must use a multi-user type of approach because we simply do not know who is going to need the data or in what form they will need it a decade from now.

RANNESTAD—Returning to what John (Peachey), Carl (Oppenheimer), and others have said concerning the amount of data necessary, I believe Carl said, "the more data you have, the better the result you get." This, of course, is true in most cases, but it also depends on the data, its quality, and the amount available. When you consider the enormous amount of data received from today's satellites, do you think that it is necessary to save all this data and computerize it? I think we must divide the data into two groups. Some of the data is for immediate use and has, let us say, little or no value in the future and thus can or may be discarded after use. Other data of more permanent value should be compressed and saved for the future.

I think that this will be necessary in many cases because if you fill your computer with all available "raw" data, it would be impractical to use and even impossible owing to economic reasons. Thus, part of the data management function must be to make a selection of the data to be compressed and retained in a workable system. The rest of the data may be stored in archives and when the storage becomes a problem, eventually destroyed.

OPPENHEIMER—This point you make, Dr. Rannestad, supports the issue of devising a summarizing two-dimensional system for environmental data similar to a weather map.

AUSTIN—Austin's law is that whatever data you purge today is what somebody's going to need tomorrow.

FLEMING—I think we can make some kind of decision as to what sort of data to computerize. I think we ought to computerize that data that we need for analysis and synthesis and not computerize data which is inventory data, what I call telephone data. I don't want to ask for Carl Oppenheimer's telephone number by calling up the computer. It's easier to look in the telephone book. It can all be found by looking up one single name and I can get the telephone number. I don't want to know how many telephones there are in Port Aransas, or where, or the telephone number of somebody that lives at such and such an address. In other words, I don't want the information that is in the telephone book to synthesize anything; all I want is to just get a number out. And that sort of data does not have to be computerized, and that holds true of much hydrographic data. The data necessary for correlations and synthesis needs to be computerized.

HAMMERLE—The one point that has not been brought out although it seems to me we are getting very close to it is the purpose and the needs of data. To one person the need for data is obvious because he needs it and the other person doesn't want it because he doesn't need it.

What do we need data for? What is the purpose of data? If you talk to a scientist, data are for research. If you talk to a person who tries to enforce laws, data are for law enforcement.

So we go back to the functions of the agencies that produce the data. EPA, for example, produces data because they need to have data for enforcement and regulation. Other agencies produce data because there is an interest in the data. Some agencies are totally research-oriented and they produce data because they serve research.

It seems to me that we haven't gotten to the point yet where we have talked about systems which are strictly for a purpose. We might go back to Dr. Peachey's mission-oriented versus goal-oriented. And here I get back to my question . . . EPA has goal-oriented systems in the sense that they are trying to enforce laws; for example, pesticides, radiation, that type of hazardous wastes. So there are a number of systems that I believe are goal-oriented. There also are mission-oriented systems which are bibliographic-type systems. There seems to be a distinction here, and it seems to me, that it is important to bring out that distinction. Whether you computerize data, or you don't, depends upon the purpose and the need. And that gets down to the point of computerization versus non-computerization.

COLEBROOK—I think that the center with which I am associated has quite a number of data systems. It started with a definite goal to help the fishermen with the idea that where there are more copepods, there may be more fish. So we went out and surveyed the distribution of copepods and told the fishermen where the copepods were, and it worked. But then somebody invented sonar which is a very much better and much quicker way of finding fish than counting copepods. And at that time, the future of the copepod survey was very much in question, and I think it was only some very, very quick thinking and quick talking on the part of my Director that kept the survey going. And now, after the survey has been going for something like 30 years, we have a 26-year time series largely by accident. We have a 26-year time series on 30 vital, non-exploited populations of organisms. And the true value of that survey is now beginning to be realized. It was largely an historical accident, but we've got it.

So, as you say, the need for historical data is unpredictable. You don't know what you are going to need in 10, 20, or 30 years time, but somebody has to guess somewhere, I suppose. We cannot afford to leave it totally to chance. I just wonder who is going to do the guessing. I think governments aren't, but I'm not quite sure who is. And having guessed, are they going to have the power to set these programs up?

HAMMERLE—I made a point this morning that differentiated between data collectors and data users and I would like to go back to that again in this same context and re-emphasize that there is a difference between the two. People don't usually go about collecting uncalled-for data. However, I know from our own experience that we do a users survey every year. Each year we try to identify existing and potential data users and these people drop in and out of the picture with time. New ones appear every year that we never anticipated before and invariably our largest criticism comes from this year's user who said last year he was not a user. This happens constantly, and I see no way around the problem.

The data manager has a thankless job and has to have a thick skin. He has to anticipate and be like a doctor—prescribe the treatment, don't let the patient tell you that he is sick or not sick, but tell him what he needs to know. Sooner or later he develops the capability of being able to ascertain from what their mission is, what their needs are going to be even if they don't realize it. At least that's what we have done and I think that it has been relatively successful.

AUSTIN—I love to make generalizations because then I can be shot down. I would say the data user normally doesn't know what he or she wants. How often have we had calls of "would you please give us copies of all of the data you have" for a certain thing. This may involve 60, 70, 80, or 100,000 observations, and you say "well now!"

As a matter of fact, we have a standard instruction the people in our service units follow when people ask for "all" data on a particular subject, no matter whether the request comes by phone, cable, or letter. Our information people have to get hold of the requester and ask, "May I discuss this request with you?" I remember a request not too long ago for all data associated with areas of upwelling of the oceans of the world. Well, that's rather a fabulous request. When we finally pinned it down, the individual was perfectly happy with computer generated vertical sections during the peak of upwelling like the one off the West Coast, and during the period when the winds were such that there was no upwelling. That's all he wanted. But had we sent him something like 200,000 stations on the computer listings, he would have gone out of his mind trying to use them.

So I would say that very often people know they have a problem and know they need data, but really don't know what they need until people who are concerned with data management lead them by the hand.

HAMMERLE—This is exactly the same situation that we have. We call it negotiating. We rarely provide the requesters with exactly what they ask for; we always negotiate to try to find what they really need.

AUSTIN—Very nicely put, I've got to remember that.

RAUSCHUBER—I would like to take this opportunity to acquaint the members of the Conference with the procedure which Texas follows for handling the vast amount of natural resources information which various state and federal agencies collect within the state.

In June of 1972, the Water Oriented Data Programs Section (WODPS) of the Inter-agency Council on Natural Resources and the Environment recommended that the State of Texas pursue the development of a Natural Resources Information System (NRIS) to facilitate the fulfillment of the specific statutory responsibilities and administrative needs of the various agencies involved in planning, developing, operating, managing, conserving, and protecting the natural resources of the State. A major objective of the NRIS would be to provide maximum availability of natural resources data and information consistent with cost and efficiency.

The first effort of the NRIS was to "define" natural resources by developing a set of categories and subcategories which would adequately "contain" the data to be later identified. These categories and subcategories would be subject to revision as the identification activity proceeded. From the outset, it was decided that the intent of the identification activity was to identify all data and information holdings which might even remotely be considered natural resources.

Table I lists the categories and subcategories that have evolved from the identification activity and represents a "definition" of the data and information which might ultimately be included in the Texas Natural Resources Information System.

The Texas Water Oriented Data Bank (TWODB) is being implemented by the WODPS with primary staff support from the Texas Water Development Board and serves as a pilot project and nucleus of the NRIS. The primary purpose of the TWODB is to provide the means to efficiently serve the data storage and retrieval, data presentation, and limited computational needs of the various participating state and federal entities requiring water-oriented data and related information. Data and information are also being made available to the general public upon request.

Under its service-oriented operating concept, the Texas Water Oriented Data Bank can be labeled as a "User-Oriented Data Bank." The process of data and information retrieval usually begins with a request from a participating entity, who may require data from one or all of the six data and information categories.

TABLE I. Coastal Data System Outline
of the Texas Water Development Board

I. Geographical Base Data

II. Meteorological Resources

    A. Climatological
    B. Air Quality
    C. Man's Activities

III. Biological Resources

    A. Animal
    B. Plant
    C. Micro-Organisms
    D. Man's Activities

IV. Water Resources

    A. Surface
    B. Subsurface
    C. Man's Activities

V. Geologic Resources

    A. Surface
    B. Subsurface
    C. Man's Activities

VI. Socio-Economic Resources

    A. Social
    B. Economic
    C. Commerce
    D. Government
    E. Archaeologic

When the request for data or information is received, it is logged in and processing is begun immediately. Some requests may be fulfilled within a matter of minutes; others take a longer time. Users who prefer much faster retrieval may choose to install a remote terminal in their own offices, which will offer them immediate access to many of the files of the TWODB. Whichever method is chosen by the user, it is the goal of the TWODB to offer participating users access to a maximum amount of high-quality information in the most meaningful form within a minimum amount of time.

KOHNKE—One of the most serious problems data centers are faced with is the estimation of what the future requirements of the users will be. After the experience the NODC's made in the past years, it seems to be useless, from the cost/benefit point of view, to archive all the data which have been originally collected. Since a couple of years, instruments are being used in oceanography which have extremely high sampling (recording) rates (in the order of a few seconds). Is it reasonable, in any case, to store the original set of data or should the total amount of the collected data be compressed in a way that the significant features are preserved, but insignificant values are deleted after well-defined criteria? For various reasons I don't believe that it is useful. The variability in the marine environment is so great that measured values have mostly to be regarded as random.

The number of requests for the original records of such high-resolution instruments is extremely low compared with the total number of retrievals. Therefore, some NODC's developed computer programs by which the total amount of the recorded values is reduced. Only significant points of the record are kept. It has to be stressed that the original recordings are not lost; by means of inventories held at the NODC's it is possible to refer customers interested in the fine-structure of a record to the originator.

With your permission, Mr. Chairman, I would like to give a simple example of such a data compression. Figures 30 and 31 show two pairs of curves: the vertical profiles of the water temperature and salinity obtained by a so-called STD device (salinity, temperature, depth). The left-hand curve of each pair is the originally observed vertical distribution recorded at 231 different water levels. A special computer program developed in the German Oceanographic Data Center reduced this set of data to 91 water levels only.

The right-hand curve of each pair is the temperature and the salinity distribution, respectively, drawn with these 91 selected flexure points. Please note that the 91 values are still originally observed data; they have not been altered by averaging procedures. The maximum (not the standard) deviation of a neglected value from a linearly interpolated one of the reduced data set for the same depth is 0.2°C for the temperature profile and ± 0.02 parts per thousand for the salinity curve.

As you see, the main features of the two profiles were kept; even the fine-structure was preserved in the reduced curve. The reduction can be carried out as far as you want. Therefore, I am now giving a second example where exactly the same data were used as input. But with a permitted error of ± 0.5°C and ± 0.05 parts per thousand, practically all the fine-structures of the curves were gone.

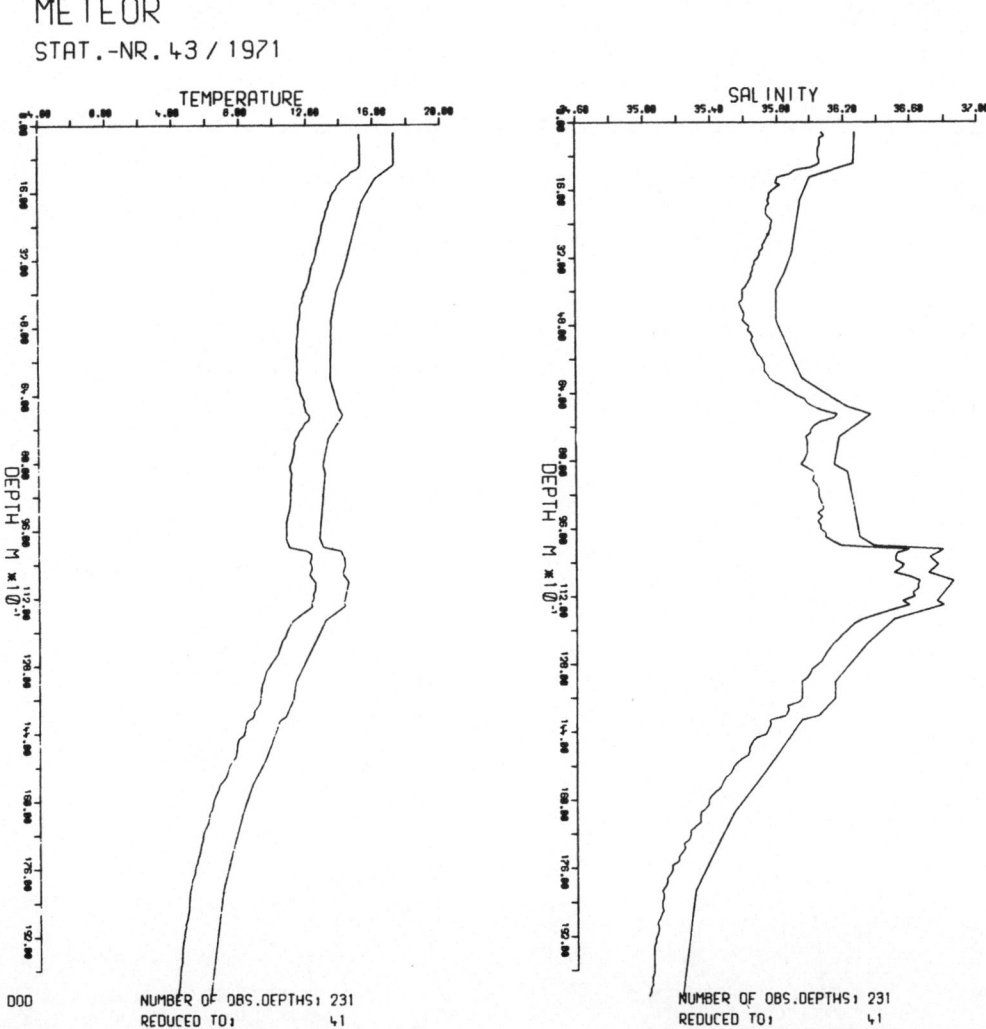

FIGURE 30. Salinity-temperature profiles.

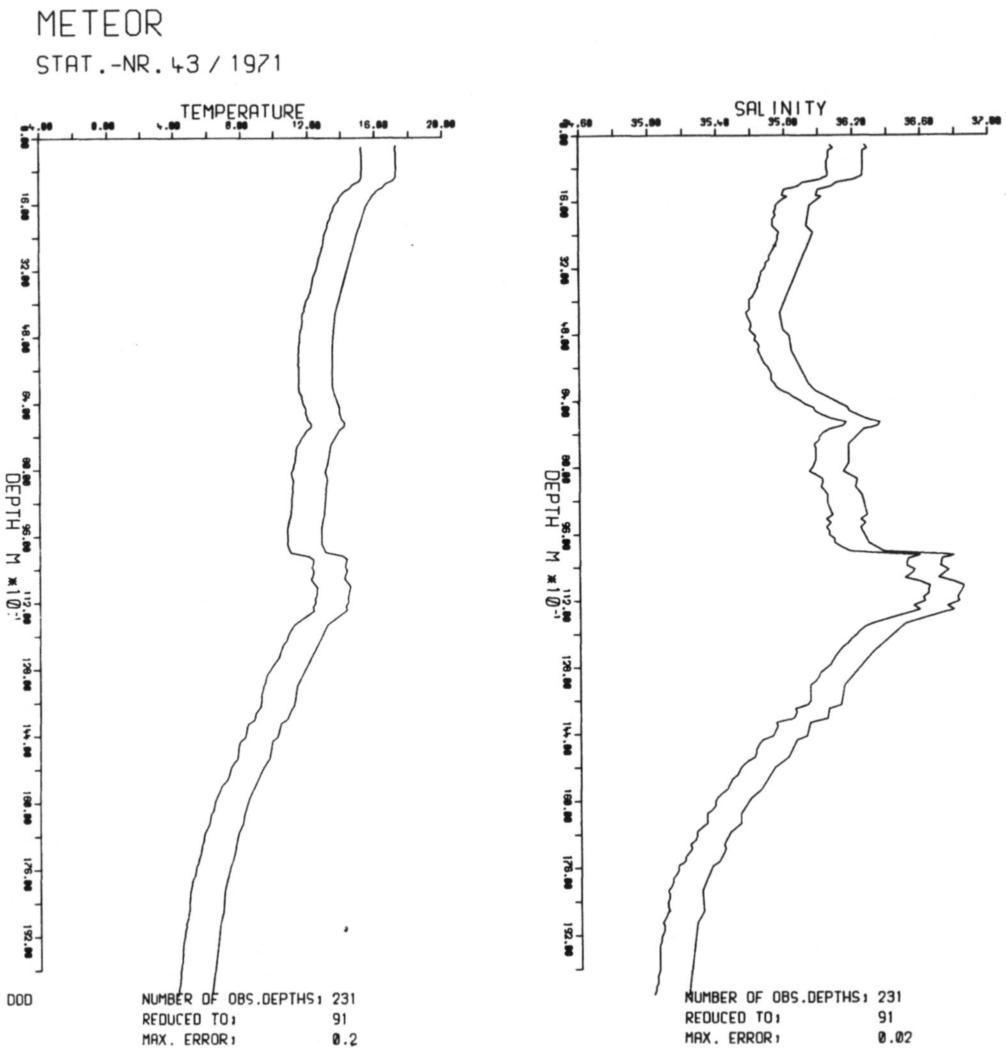

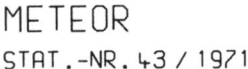

FIGURE 31. Salinity-temperature profiles.

OPPENHEIMER—If you put the information on microfiche and combine storage with computer retrieval and the ability to digitize, then the information will be compressed even further.

KOHNKE—That's true. If you have to copy a low number of microfiches just to give the requester the opportunity to have a quick look at the data, then it works fine. But if he is going to peruse large volumes of data, then it would be inappropriate to have it on microfilm or microfiche. Either the data center or the customer would probably have to digitize the analog records.

OPPENHEIMER—Can't you compare the cost of digitizing the data with compressing it the way you have done? It might turn out that it will be cheaper to put the information on microfiche and then redigitize it when it is called for.

KOHNKE—The risk is that you may have to redigitize one profile, let's say ten times per month. This would be extremely expensive and time-consuming, too.

OPPENHEIMER—The technology to do that could be developed.

NELSON—Whose responsibility is the cost, the user or the data bank? Kohnke is putting it on the user, maybe rightly so.

CUTBILL—The general point that seems to be coming out here is data gets used at all kinds of levels and in all kinds and degrees of synthesis and it is an organizational management problem to make sure that all the requirements at all the levels are met. I was wondering whether Miss Scott would say something about this because this is something that the Biological Records Center has been very concerned with, the different levels in which you wish to hold information.

SCOTT—May I show three slides to demonstrate this? We are working on a three-level hierarchy now at the Biological Records Center. I started to talk about this when we were discussing ways we see the museums fitting in environmental data systems. What we are trying to do is this: the Biological Records Center as the national center has local centers collecting detailed information about the environment in their area that is stored in some detail locally, for example, in a museum. This data is condensed and fed into a national center where it is stored. The data is, in turn, condensed and fed into an international network. Such national centers are being set up all over Europe. I think at the last count there were nine countries that were setting up a records center and we are trying to get more collecting data.

We are concerned with collecting data in the form of the presence of species in grid squares. When we started we were just mapping on a national scale and we wanted the best size of square which would give a good picture of the country-wide distribution of the organism that didn't coarsen the data so much that you lost the distribution, and also that didn't give more data than we could possibly handle. We found that for a nation-wide map the 10 km. square was the ideal unit.

Then when we started encouraging people to map in more detail and collect information in more detail locally, we found that the 2 km. square (or possibly a 1 km. square for a small county) works well. Figure 32 shows the distribution of certain plants in one English county on a 2 km. square grid. Figure 33 is an example

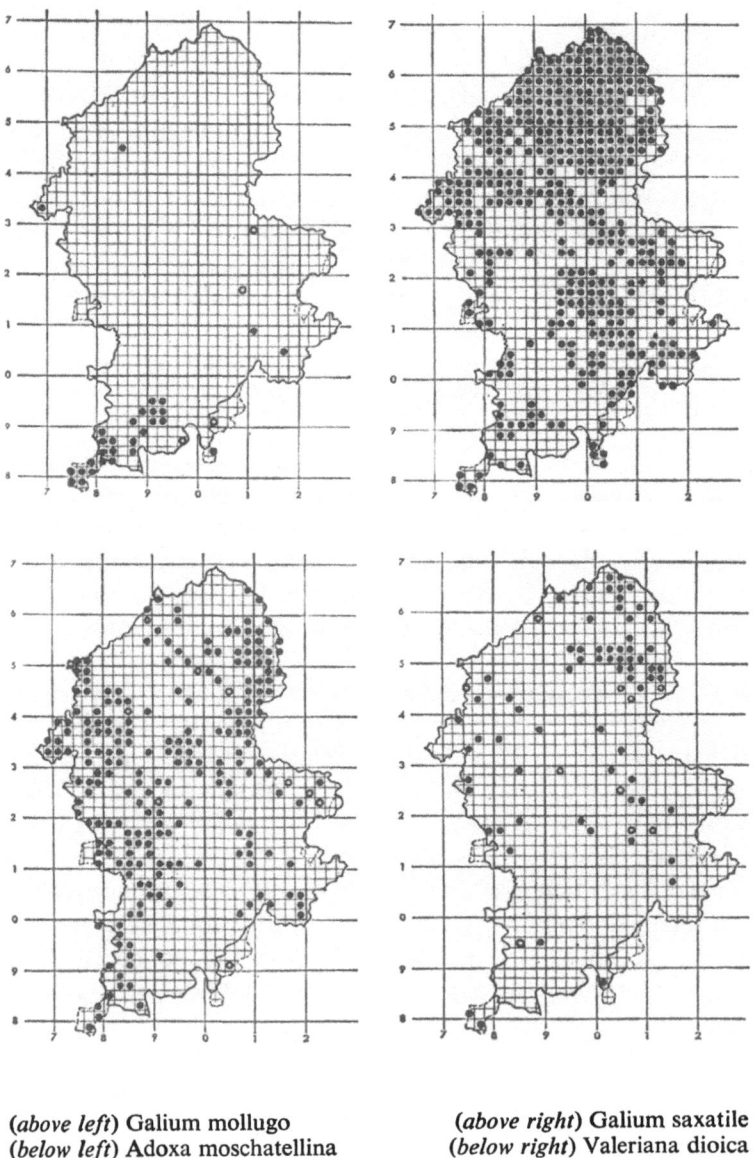

(*above left*) Galium mollugo   (*above right*) Galium saxatile
(*below left*) Adoxa moschatellina   (*below right*) Valeriana dioica

FIGURE 32. Distribution of certain plants in one English County on a 2 km. square grid.

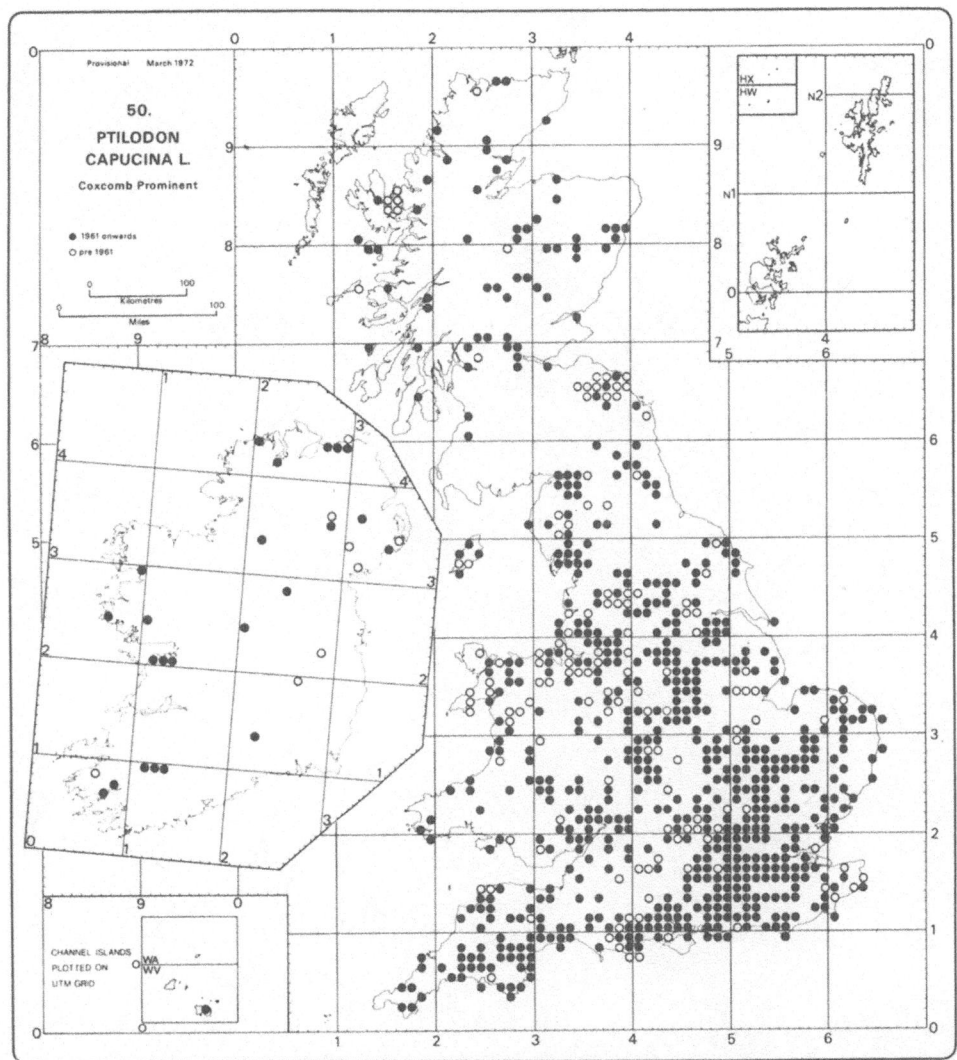

FIGURE 33. Distribution of a plant species for the British Isles on 10 km. squares.

of one of our maps for the whole of the British Isles with 10 km. squares. The county in the previous map is an area of about 20 or 30 of these squares, and obviously, if you try biologically mapping one county on its own with that number of squares you get no clear distribution picture at all. If you reduce it a factor of five then you get a good picture. As we apply the approach to Europe, we increase the scale, again by a factor of five, and map on 50 km. squares (Figure 34).

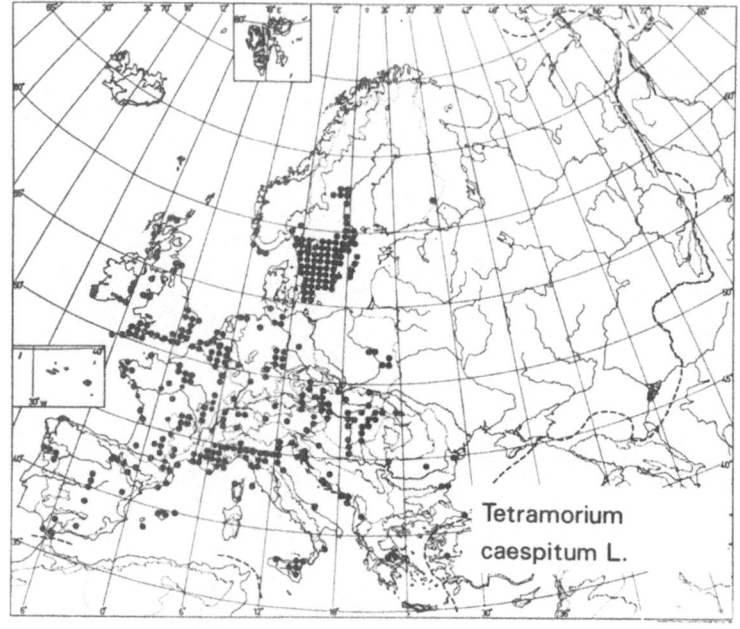

Original by Dr. Ch. Gaspar, Fac. Sci. Agr., Gembloux Belgium

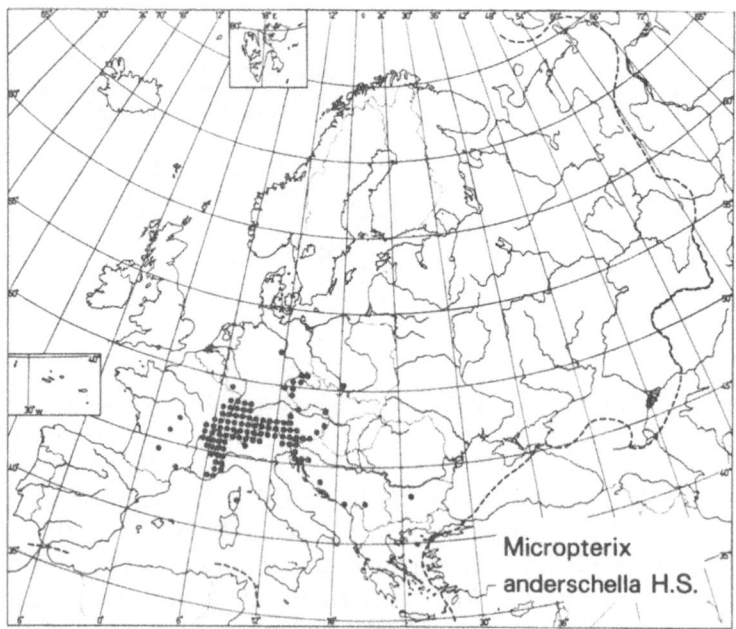

Original by J. Heath, Biol. Rec. Centre, England.

FIGURE 34. Examples of 50 km. square distribution maps for Europe.

It seems to me that this sort of hierarchical management system is one that we could be thinking about. You get condensation of data at each level, and in an international data center you have a certain subset of the available data. You can refer back whenever you want to, to the next level; i.e., to the national center, and then again the national center can refer back to the local center.

HELMS—May I ask a question here? How do you store original data? I understand they take the occurrences of certain species, I don't know as I'm not a biologist, but of a flower or so and you publish this automatically so that people can go up and see the Old Man's Beard (that is a flower?) were to be found here and there in this or that year. But do you also store the original data that the Old Man's Beard were found in this or that particular field belonging to Mr. Jones? How do you describe the site?

SCOTT—Here again, we do have a multi-level management system. The information that went into the making of the national map is stored in the central computer data bank simply as a list of numerical grid references and species codes. Going backwards from that, we have more detailed information; for example, the occurrence of Old Man's Beard in a particular field; some of that will be stored in slightly more detail on punched cards. Then going back another level we do in fact retain the true original data as it came in to us. Because we do find that in a lot of cases people want to see the original information. Whether it's because biologists are very conservative, I don't know, but a lot of people want to see the record of a particular plant in a particular locality and won't believe it's there unless they actually see the handwriting of the person who made it. In the computer we have to have a code saying where the original information can be looked up.

PEACHEY—Mr. Chairman, in a way that facility, which is common of course to many systems, is the heart of a management information system, to pick up data at the right level.

While I've got the floor I would like to go back and ask if it was seriously meant that we would not in the end computerize nearly all office, desk or bench data. Consider the effects of technological advances on a curious and hitherto effective information system which cheap copy finally buried.

The office file is, in many ways, a much superior invention to the scientific paper. It was a reactive system and originally in most western governments consisted of a folder enclosing sheets of paper on which one person wrote their views on a topic and addressed it to somebody else who then, and I think still do in Washington and Whitehall, endorsed it by ticking through his name to mean that he has seen the material and added his own comments below. No new sheets of paper were used until all original space had been used up.

The advent of cheap copying has ruined the efficacy of this system because parallel texts are distributed "sideways" and never come together again in the relevant file. And I challenge anybody here to take me to their filing system and be proud of it. And it's that, in a way, that perhaps got us all here. Farmers make bad gardeners. Perhaps really a discipline on all people who apply for money for data management is that they stylize many of their routine housekeeping functions.

But to take the point further, I would assume that escalating costs of personnel, problems of moving people into the inner-city to work, mean that a great deal of material will automatically have to be stored in low-cost computerized form for pure housekeeping reasons. You just cannot any longer get people in a capital city who can even trace a record for you manually. You can't pay them enough, they leave, or they are not secure enough, or there is a general shortage and I think that this will act as an incentive to make part of our debate obsolete.

It seems inevitable that as a society we shall and should welcome the almost total commitment to electronic means for much of what we call office work. Certainly responsible administrators admit that we have reached a crisis stage. Does the scientist perhaps feel this same pressure of paperwork?

OPPENHEIMER—Absolutely and unequivocally. This is the basic reason for this Conference.

CUTBILL—I think you have a fallacy, John. You think that once you have computerized data that all of your problems are over. I don't think your problems have anything to do with whether or not it is computerized. By and large, we have seen from Dr. Austin that there is an upward progression from manual systems to computer systems as the pressure comes on you. These are really incidental as it's the economic pressures that made you do this and the pressures of what you want to do. You won't solve your problem when you have done them. Your management problem is still there and your finding problem is still there and everything else. You know, you just saved yourself a breathing space.

SCOTT—Data preparation isn't any shorter than paper preparation is it?

PEACHEY—Our children will certainly have to learn to keyboard and to interact directly with complex on-line systems.

OPPENHEIMER—About four years ago, as a field ecologist, I came to the point where I was about ready to leave the field of ecology because of the difficulties of integrating the various parameters of physics, chemistry and biology to describe ecological conditions. The only alternative was to go to a computer-assisted system. And this is what we went ahead with. I'm not a computer person, but by teaming up with an associate, Bill Brogden, who has a command of computer techniques, I found that my frustration disappeared. There is a mass of environmental information just ready to be placed in a computerized system, or already available from computer programs. In the last two or three weeks we have answered questions about our environment that there would have been no hope of answering three or four years ago and there would be no hope today if we hadn't gone to a computer-assisted type of arrangement.

This was the whole purpose of my efforts to bring this group together. Because again, as a working field ecologist, I agree with Peachey that there is no hope for us in the future unless we go to a complete and dedicated computer-assisted data system.

GORE—Speaking as a user, not a data collector or banker, I see that the rate of accumulation is very rapid and that unless we or those mainly concerned with the storage of data do not resolve the question of throwing away data in a rational and sensible way, some discarding will be forced upon them in an irrational way by the circumstances of growth. I don't see any alternative. I don't see whether it's computerized or not computerized makes any difference at all, it's merely a matter of time. So I think that we might as well look at the kind of things that we really do jointly consider worth keeping and not keeping.

AUSTIN—It's interesting, though, how man's inventiveness keeps us maybe a little behind or a little ahead—I'm never really quite sure which. We can now, to an increasing degree, use laser technology and we now speak about storage in terms of trillion bits. What the upper level is, I haven't any idea.

GORE—Will you be able to use data when it's compacted into a state of very high density material?

AUSTIN—Oh, sure.

BROGDEN—I don't think we are talking about throwing away data but maybe just throwing it back to the next lower level of hierarchy. While I may not want data at a primary level of access, I may want it back at the basement level of access.

GORE—It will cost a great deal of money to do that.

CUTBILL—If I am suddenly smitten with an interest in the oceans, heaven forbid, I am going to want a data filtering mechanism, an intellectual mechanism which indicates the kind of an organizational problem that we have.

AUSTIN—I think if we have done nothing else here today, we have highlighted the valid position that there is a variety of data users. Mr. Fleming's comment about the telephone is absolutely right. But if I manufactured instruments, telephone instruments, and I was interested in some aspect of how many calls are made per unit of time because I have a new type of an instrument, I would be very keenly interested in information that doesn't interest him at all, because he isn't interested in making telephones.

PEACHEY—A telephone directory is generally compiled and set by computer in the first place.

OPPENHEIMER—I agree with . Tommy Austin. I think the maximum readable computer speed is the speed of light, it's just a matter of achieving it.

AUSTIN—We are almost there now.

ROSENFELD—If you want it.

AUSTIN—There are those who do want it, there are those who have to have it. It's not just a matter of whether they want it or not, there are those who need it.

CUTBILL—Before we get to the end of this session, can I just switch and focus in on one other thing that was mentioned but not dealt with and that is the possibility that data is collected for which there is no immediate purpose. There is a general feeling that data is collected for a purpose in the context in which it is going to be used. We've got a pretty wide spectrum of people here with a lot of contacts. I would just like to ask who is involved with an organization that deliberately, as a matter of policy, puts a percentage of resource into collecting data for which they have no immediate purpose. I have been involved with a geology group that has done this for several years and for us it has paid off no end. We are now sending all that wonderful data that we didn't want to the oil companies.

AUSTIN—Sure, in Oceanography.

CUTBILL—But is there any conscious policy around in organizations like this at the moment?

AUSTIN—I think so, as it originated in ship's logs from which Maury originated his treatise on ocean elements. In Oceanography it's called "ships of opportunity." When a ship leaves port and heads from A to B, the crew may or may not collect data every watch. We, in 1954, began collecting sea surface temperature and salinity samples between Hawaii and Tahiti on the Matson liner every watch. In 1958, there was a major climatic change in the equatorial Pacific Ocean. Luckily, even though we started the collection at a time when we really weren't sure exactly what we were going to use it for, we thought that as the years passed it would have more and more of an application in the fisheries. It sure did. We can now predict with a reasonable degree of certainty what is happening in the fisheries.

ROSENFELD—There is another reason, and that is the small incremental cost of getting other data on a very expensive expedition. That ship or satellite is operating at a very high cost and one more instrument put on by someone for speculation is a tiny incremental addition to the cost of data collection.

AUSTIN—Up to a point, there must be hundreds of thousands of plankton samples stowed away in the closet just because it's so easy to tow a net behind the ship.

ROSENFELD—That is precisely the point we have been talking about—easy and cheap data collection for no obvious future use, followed by costly storage and management, and then, finally, a use, maybe.

CUTBILL—Should we be trying to focus in on any general policy ideas here or that sort of thing, or does the randomness of individual things actually dismiss it?

HAMMERLE—I think this is a thing that comes of experience in this collection of data that nobody needs. I think you can come about it in two ways: one in which it is the policy of your organization to do that; or, two, in which a person with sufficient authority to authorize it to be done just has a feeling that it may be necessary. He may later on turn out to look either very stupid because he did authorize that or he may be the savior because he did. It could turn out either way. I don't think there is any specific guideline other than experience that would indicate what you ought to do or not do.

OPPENHEIMER—In other words, all managers of computer systems have to be the boss so that they won't be criticized?

HAMMERLE—Well, I think they shouldn't just be the data people. And I think this is really important, that they are not just the computer people, statisticians, they should have a lot of technical knowledge in the field of science or engineering or whatever it is that they provide service to. If they don't they aren't going to have any intuition or the practical experience that indicates what they ought to be doing.

GORE—I really think that it is very difficult to know what data to keep, and the cost aspect of it is very important. As Dr. Rosenfeld mentioned, you can gather the data as you are doing something else, then the cost is small. But the cost of gathering new data for some purpose that you have no idea whatsoever of how it will be used, is obviously not going to be done because the cost is high. And when you look at the laser technology coming up, which is extremely costly to store, it won't be done until that cost comes down. There are too many things that are so totally economic related that are intimately interwoven with this whole problem.

ROSENFELD—If we had unlimited funds right now, I don't think we could computerize all those data that Dr. Austin was showing that were non-computerized.

AUSTIN—I can't even envision what unlimited funds means.

ROSENFELD—Well, I mean you could hire every person you wanted and have all the hardware that now exists.

FLEMING—Part of the data we collect is, I think, a function of the culture we live in too.

PEACHEY—I wanted to ask a technical question which is, in a sense, related to Mr. Loudon's point earlier this morning. When we measure things we degrade the phenomena. Now I seem to remember in one ecology class we were shown something that measured climatic fluctuations based on an electrode and it really gave you a weight of silver which showed you the degree of fluctuation. I want to know whether there has been much progress with producing self-aggregating and self-analyzing representations of a concern. If you look at dot diagrams, for example, would you ever be able to get a picture so you can say that's nasty and needs watching without really having to look at what the individual measurements were or what composits they are made of. They might just be the 25 pollutants identified in the Global Environmental Monitoring System or something like that. Is there such an aggregating and impression-making capacity, and do satellites do some of the work?

CUTBILL—I think this is a little trespassing on tomorrow's session.

PEACHEY—The point I was making is that we are talking of non-computerized applications and in the back of our minds there is always this business of retrieval and compression and it seems odd to me that we are still stuck with numbers, letters and ordinary words. We have graphics, but they are still diagrams; they are not a wholly new representation of a very crushed and complicated picture.

BROGDEN—Do you think we should have invited an artist and environmental poet or something?

PEACHEY—Well, I thought that you did get painted representations?

OPPENHEIMER—We approached this somewhat analogous to the weather map system. I keep coming back to this because it has always intrigued me as an environmentalist. We teamed up two artists with biologists and chemists and actually went out and described the environment with artistic license which was supplemented by scientific documentation as a first means of identifying environmental niches. A picture combined with compressed environmental information is a way of compressing data. As Dr. Peachey said, it's like his desk copy, you put everything on one sheet and then put that sheet in your archives and have the capability of pinpointing any definite spot on your sheet to pull out the information of that spot. And this is what you are talking about in three-dimensional environmental descriptive coverage.

AUSTIN—I did that.

NELSON—Dr. Margalef's work on species diversity indices in getting the community down to a number we can evaluate, is a case in point. As we mentioned earlier this morning, we might have to develop a system where one number represents more than the biological community for models or indices, something we could analyze and store as the number we use to trace environmental trends. Of course, that number would always have backup raw data we could examine. That would be the user's responsibility. The user could check the data used to formulate the index or model and this information could be computerized if we wanted to spend the time and money for further analysis. But should we computerize every bit of information? We don't. It should be the user's responsibility, possibly, or the user might convince some institution to computerize the information required for problem solving.

ROSENFELD—Now we are back to information, not data. If you have six different biologists they will want six different indices.

NELSON—We are using twelve.

ROSENFELD—Well, nevertheless, as Dr. Loudon said this morning, we are now back to the information versus data discussion. We did make a distinction this morning and we should probably stick by it. We have now, just in this last five minutes, gotten into interpretation of information as data.

CUTBILL—Can I just lay to rest the information versus data one please. This was done for me years ago when I started computing by Dr. Robinson of the Canadian Geological Survey who pointed out that the surveyor's well position is his information and my data.

LOUDON—I am surprised by this discussion about how one condenses non-computerized data, because it seems to me that this is what science is all about, the function of science. We attempt to condense the very large amount of observational data by converting it into a more compact and communicable form of geological knowledge by synthesis and interpretation. The model is a compact representation of the crucial features and relationships which exist in the data. Within the computer also, the model may prove the most compact means of storing the patterns observed or predicted within sets of data.

CUTBILL—Yes, I am certain of it, but you will admit that this is a repeating process and the information out of one stage overcomes the data input from the next stage or the thing would never work at all. You would have to do everything from scratch.

LOUDON—I don't think I would.

PEACHEY—Mr. Chairman, we haven't cleared up the problem of vocabulary. I have noticed that many of us accept no difference between the two terms—data and information.

ROSENFELD—I hoped to make a distinction this morning for the sake of consistency in this week's Conference. I have heard people define information as everything; it includes data and interpretations; it's really not very important. The point of the level of condensation that we are talking about is that if you lose the raw data on which information was derived then you are now accepting as fact a single scientific interpretation of which there may be many from the same data.

OPPENHEIMER—I hope that we haven't considered replacing data with interpretations. The current discussion considers the thought that compressibility and condensibility are interpretations. I do not agree with that analogy. I believe that the original data bit can be condensed to a form that is very easily put in a system that may not be the same as interpretation.

CUTBILL—I believe we have reached the time to convene. I am sure that after this stimulating exchange of ideas today, we will have plenty to discuss during the remainder of this Conference.

IV. COMPUTER DEMONSTRATION SESSION

Session Chairman
Dr. Carl H. Oppenheimer
The University of Texas
U. S. A.

Principal Organizer
Dr. William Brogden
The University of Texas
U. S. A.

"Computer Demonstration Summary"

Dr. William Brogden

This stage of the Conference was devoted to displays and demonstrations of environmental data systems. We were fortunate in having demonstrations which covered a variety of applications of interest to both environmental scientists and managers. The organizers would like to take this opportunity to again thank those participants who provided displays and demonstrations.

Some of the capabilities of the Texas Water Development Board's "Texas Water Oriented Data Bank" were illustrated by both on-line demonstrations and displays. Using a telephone link to the TWDB computer in Austin, Mike Ellis and Don Rauschuber demonstrated the use of the interactive "MONITOR" program to locate data sets of interest to potential users from the large number of meteorological, biological, hydrological, geological and socio-economic files available in this data bank. Other capabilities of this system were illustrated by displays of plots, maps and statistical summaries.

Dr. Russell C. Eberhart of the Johns Hopkins University Applied Physics Laboratory demonstrated the "Research and Management Shoreline Data Bank." This interactive data bank, using the IBM Generalized Information System contains information on permits requested and granted for construction affecting the shoreline of Chesapeake Bay.

Application of the Battelle Automated Search Information System (BASIS) to bibliographic data was demonstrated on-line by Dr. G. J. Kovacs. This generalized system has been applied to a number of environment-related data bases.

Dr. William Brogden of the University of Texas Marine Science Institute demonstrated an interactive system for the retrieval of life history information pertaining to organisms of the coastal zone using ENVIR. This system was designed to support research on coastal zone management.

A display on the computerized data management system for the "Flora of Vera Cruz" project was presented by Juan Toledo M. of the University of Mexico Biological Sciences Department. This system is designed to provide retrieval of data in a variety of forms such as distribution maps, listings, etc.

DISCUSSION

OPPENHEIMER—The Conference is now open to a discussion of the computer remote capability, their ability to inter-relate, the possibilities of creating a network of environmental data bases, and what we should look for in the future in the way of recommendations for the last day of our session. We must look to the future to try to anticipate problems that we will have with interfacing information systems. Perhaps we could start out by asking some of the users, like Dr. Kovacs. What is the possibility of interfacing your system with some of the others? This can be approached from an economic basis, from a proprietary basis or from a scientific basis.

KOVACS—Well, by interfacing, do you mean interfacing a machine, groups of machines, or essentially user groups? Perhaps I could just give a very quick scenario of the kind of application that was demonstrated here. As many of you probably know, a number of very large scientific and technical oriented data bases such as Chemical Abstracts, The National Technical Information Service File, and there are a number of others, such as ERIC (Educational Research Information Center, and Lockheed's PANDEX and TRANSDEX are offered on-line today. And as a matter of fact a number of others such as Engineering Index, which until several months ago, were accessible exclusively through batch tape searching mechanisms, are now being offered by Systems Development Corporation on-line.

So there is a very strong trend for these kinds of general interest bibliographic orientated data bases to be offered to large public groups on a time-shared basis. I believe at last count the Environmental Protection Agency, at least the Library of the National Environmental Research Center in Cincinnati, is tying on-line into something like 15 bibliographic data bases.

Obviously, one reason for this is because of the economics of the situation. We have a situation today whereby one can access one of these files containing anywhere from 200,000 to a million records of information for approximately the cost of $35 to $60 per hour of terminal connect time. While this cost may sound expensive, when you consider that in a matter of 15 minutes you can run a fairly complex search (one that would simply be impossible to conduct manually in a practical way), you are getting an awful lot of value for that $30 to $60 an hour at the terminal.

Another factor dealing with economics is that obviously not everyone can load these files up on their own system. In effect, we see a situation where two or three organizations are establishing these large files, maintaining them on a full-time basis on disc storage, and offering larger user audiences the service of accessing these files. I think the best example of this is the National Library of Medicine MEDLINE System, which at last count, well I don't know what the precise figures are, but an enormous amount of hours per month are used in searching that particular system.

So, in summary, we are seeing a growing trend of textual, bibliographic, call it whatever you like, type information systems being offered from various points in the country, either by government or by private industry to the general public for costs that are very reasonable. An example of these services, Figure 35, shows Battelle's ties to data bases pertaining to energy data. We should mention that the cost of a typical, portable terminal is another major factor for the increased usage of systems such as the kind I described. Terminals rent upwards from $30 a month. Finally, the ease of use of these systems, which is another subject all to itself, is still another reason for the general appeal and heavy usage that has evolved.

OPPENHEIMER—What about data other than bibliographical?

KOVACS—O.K., well that, I think, is a totally different area. Data doesn't have the characteristic of bibliographic references in that the latter has general appeal. Data tends to have much more localized interest than does, say, the chemical literature, the engineering literature, or the medical literature. So I think, and perhaps the people from NOAA might be in a better position to respond to this, that it may be some time before we see the general public, and by that I use the term very loosely, and refer to a scientist in a particular discipline, tying into mass-produced data oriented files. I think this is off in the future to some extent. In fact, you mentioned the business of tying one computer into another and sharing data between systems, perhaps even sharing programs between systems.

I think the most futuristic development work in this area, which some of you are aware of, obviously, is the ARPA network, wherein some 45 computer nodes are networked together by high-speed, 50 kilobit per second, lines, allowing data sharing and a fairly low level of resource sharing, wherein a job that is run by any particular user is routed to the computer best equipped to run that job. Data files are stored in one place and a user can access that data file from his own terminal through special interfaces called IMPS or TIPS.

The BEIC Network

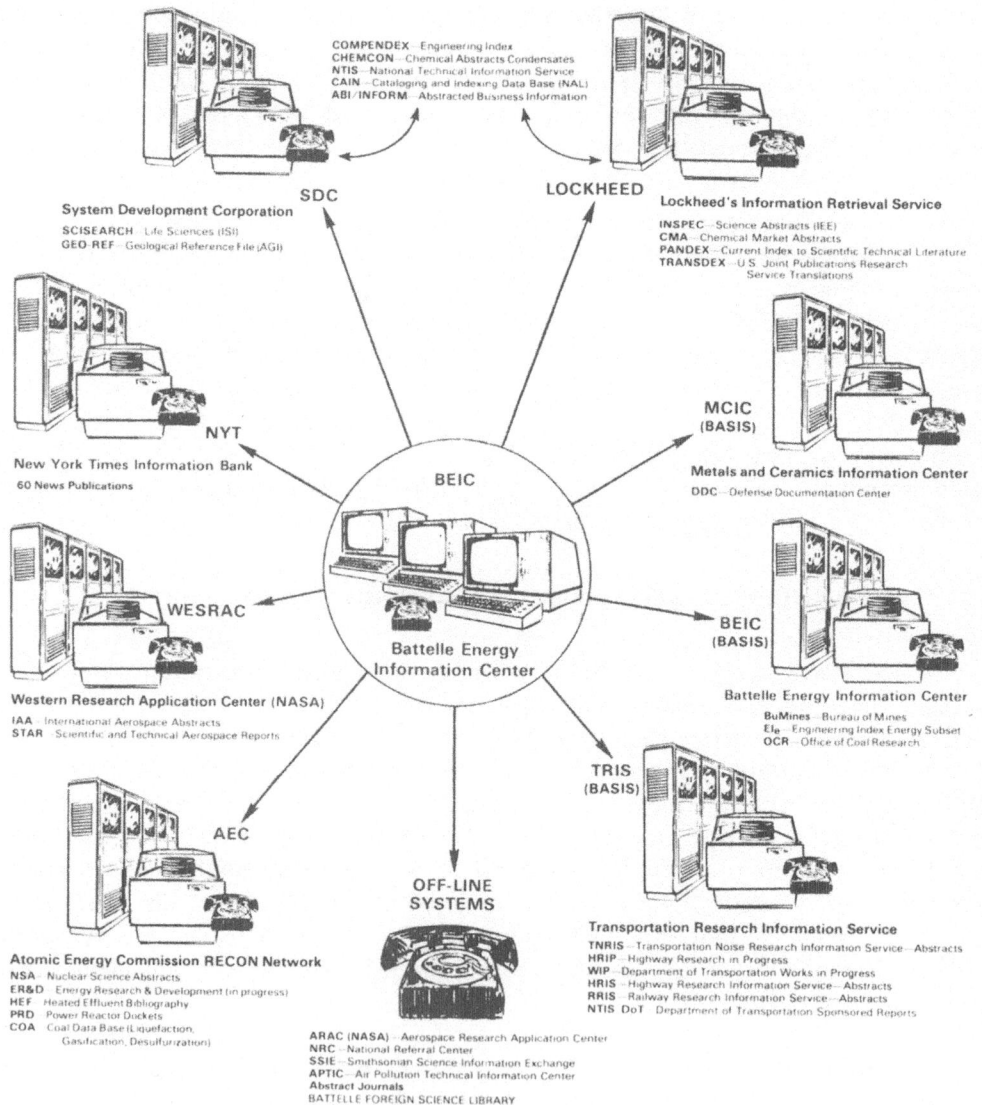

FIGURE 35.   The Battelle Energy Information Center Network.

Another example of that is the TYMNET network operated by Tymshare, Inc. TYMNET was set up originally by Tymshare for their own sharing service, but they now have offered telecommunications facilities to independent groups such as Battelle. We can tie our own computers into the network and have the benefits of reduced communication costs through economies of scale. For example, to tie into Battelle's computer from here in Houston, instead of calling Columbus, Ohio, directly, I dial a TYMNET number here in Houston and I get routed automatically to the host computer, a CDC 6400 in Battelle. This is a value-added type network wherein Tymshare essentially leases AT&T lines, adds a number of features to those lines such as error detection, line conditions, switching, etc., and enables the user to get by with a telecommunications cost about 30-60 per cent cheaper. In addition, the network allows the user to communicate with virtually any type of terminal device, which eliminates the frustrating problem of compatability.

OPPENHEIMER—Jim (Noel), can you estimate what percentage of the total oceanic data available in the United States is in the NODC files?

NOEL—Well, in terms of station data for the open oceans, I think the figure we have been using is about 80 per cent.

OPPENHEIMER—Eighty per cent, how close to land are data available?

NOEL—Just about as close as someone can throw a Nansen bottle over the side. We have heard the question of what is coastal and what is open ocean. On this sort of thing you really can't draw an exact line. If somebody comes in and takes a sample or a station near the shore, we put it in the files, we don't throw it out. But we have not set up a coastal category of nearshore stations.

BERG—I would like to make a comment about the Advanced Research Projects Agency Network (ARPANET). ARPA is part of the Defense Department and the network is completely supported by ARPA. Theoretically, one should be able to route jobs readily from one computer to another on the ARPANET, but in actuality that is not yet really a standard operation. As a matter of fact, due to the experimental nature of this network, many such operations are not yet available in the ARPA network. You can have jobs run on computers in other locations, but you have to make arrangements in advance on an individual basis, and there is a good deal of protocol that is required to make the routing work. Computer-communications networks are obviously the next advance in computing, but they are just emerging from the experimental stage. The cost figures for network operation are something that everybody argues about and no one knows exactly what the true cost figures really are, particularly for more complex functions.

KOVACS—If I may interject one point right here? It is true that the cost figures that one reads about or hears depend on who's repeating them and what kind of vested interests they have. But I think one indication of favorable economics is the fact that the telecommunications concept developed as part of the ARPA network is being, or is in the process of being, applied commercially through groups such as Telenet Corporation and Packet Communications, Inc. The latter are borrowing ARPA technology and developing their own telecommunication networks based on that technology and offering them on a commercial basis. These new ventures have published rates as low as $1.25 per kilopacket transmitted, completely independent of distance. The user will be charged on the basis of how much information he transmits rather than how far he transmits it. This represents radically new thinking and is receiving enthusiastic response from users.

BERG—Some companies are developing these networks, but how long they are going to be in existence and what possible costs are going to be incurred are, I think, unknown at this point. Network companies are being started in the same way that people start in many businesses and whether it's a viable concern isn't, of course, known yet.

Other types of systems besides bibliographic and data systems are advantageous to use in a computer network mode. I am aware that EPA has at least one of these systems on a network. A very useful type of system is one which allows access to modeling capabilities. The use of such a system in a network would allow one to remotely access a model someone else has in his computer, to enter your data into it and retrieve the results from the model for possible further processing at another computer. Additionally, there are methodology types of systems from which you could obtain algorithms and methodologies for use in a remote mode of operation. As far as the data systems are concerned, STORET is one that is accessed by users around the country in a time-sharing mode of operation. ENVIRON is used rather like a network as a way of accessing different systems having not only bibliographic information, but monitoring data itself.

OPPENHEIMER—Are you on-line with either STORET or ENVIRON?

BERG—We are not, but we are a part of the ARPA network. To get into STORET you have to have an arrangement with EPA, or be a contractor for them. STORET is, as I understand it, and perhaps Dr. Hammerle can clarify this, limited to use by the EPA organizations, the States and other governmental organizations and also some private contractors.

HELMS—I am particularly pleased about the comments on the ARPA network and I can't repeat them but I think that this is one of the examples where we computer people more or less have told the rest of mankind that we master all technologies and, very shortly, we can do everything for you. What we seem to have forgotten, some of us who were enthusiastic about and still are enthusiastic about computer networks, is that we suggested a research development venture and not a service venture. And many of these networks which you hear about are really important research and development ventures and only barely have they reached a serviceable level including service for a reasonable cost.

There is one other point which I should like to dwell upon. It has been mentioned lately that you can share models throughout a certain network. What you can share, in a way, is other people's capabilities, and I think this is one of the most important aspects of computing networks; it is not these massive transports of data—I predict you won't do that. But you would like to share other people's capabilities. In this respect, this is one thing which we haven't heard about in the Conference.

I am not an expert in your field; I am just one of these computer guys who trys to give my modest advice to some other people; but, I can tell you one thing which I have done at home where I have given computerized advice to some sociologists. You see, we have in my country rather centralized files for all people in the country. I won't discuss all the social implications of that, but the fact is, that if you have this kind of files in a country which has about five million inhabitants, these files are used for many purposes and they are also, of course, very useful for sociological research. Now it was clearly realized that although we are only five million Danes, to make sensitive sociological research and to browse through all these files and records time after time made no sense. What we have developed, we have been an advisor, is what we call the Median Population Register of the population. This means a statistical sample of the population, which in fact has proved very effective for scientific purposes, and it has been done also in such a way that it includes time-costing effects.

I am just wondering if things like that might not be one of the solutions to your problems. You are not interested, I presume, in every single observation. What you are interested in, still I presume, must be to observe the trends. Isn't it possible for you to make this kind of mini-population registers and files?

KITSCHLER—I have a question to Dr. Helms. Do you have a kind of data collection or data production in Denmark too, considering the fact that you collect a lot of information about areas of the country which might be used for purposes which might not be evident? We have a problem in the Federal Republic to carry a data collection at all because this could be a dangerous problem.

HELMS—It certainly is a dangerous development and I am going to say that beside the general privacy act we have in my country, we also have more special legislations for data protection. However, these centralized files containing information about the population are protected by special vaults underground which means that there is now a set of rules which applies to the files.

OPPENHEIMER—Gentlemen, I am afraid that our schedule does not permit any further discussion at this time. We will have a short coffee break and then move on to the next session.

V. STATISTICAL THEORY APPROPRIATE TO ENVIRONMENTAL DATA

Session Chairman
Dr. Carl H. Oppenheimer
The University of Texas
U. S. A.

Principal Speaker
M. Lenco
Ht Comite' de l'Environment
France

OPPENHEIMER—It is time to hear from our colleague from France, M. Lenco, who is taking Dr. Theys place. He will speak on statistical theory appropriate to environmental data. I understand that his talk is a little bit different than Dr. Theys proposed.

"Statistical Studies to be Initiated in the Field of Environmental Sciences," (Sponsor of the Interministerial Group on Environment: "Statistics, Accounting and Economical Concepts.")

M. Lenco

Introduction. Owing to the nature itself of environmental statistics, the environment is not a mere new concern but implies that a new dimension must be taken into account among other classical fields and associated information systems.

Besides the centralization and exploiting of data, that are, indeed, a result of administrative activities, development of environmental statistics requires, consequently, some modifications, details or design of classification, and also, that sections of forms must be changed or extended. This would, in turn, produce new tables in the existing statistical information systems. The number of new specific operations seems to be limited.

One task linked to the development of environmental statistics, therefore, consists of offering complementary requirements to various services that produce data, and maintaining a good level of co-ordination among them to get a "whole" coherent with the environment.

Data introduction and the outset of synthesis will come, as gathering of data peripheral elements of audit and the evaluating process are proceeding, as soon as their number is sufficient: stock audit material balance to finally reach ecological audits.

However, it must be mentioned that localization of environment and data requirements are still poorly known, and so are the impacts upon men of middle and long-run phenomena. Moreover, thresholds of acceptability are changing over time; consequently, one can only make a vague, rough and defined-in-time estimate of statistics that are necessary for a proper study of the environment. Further, this must be done being most of the time unsure of what kind of data are to be gathered for further uses. This is actually essential in the choice of a starting year of the statistical work.

One can distinguish several types of statistical operations:

1. Consisting of reviewing data potentially available (products of administrations, surveys, files).

2. Those that aim to integrate the environmental demand into existing surveys for completion.

3. Those consisting of preparing and initiating new specific surveys non-regular in character.

4. Those that aim to undertake statistical studies inherent to environment from existing data, or complementary data from various origins to be gathered or evaluated, from expert consultations.

We shall be mainly concerned with the latest category.

Division of the territory into study areas. In order to study the man-made environment and associated phenomena, it would be useful to design a classification of municipalities and neighborhoods for cities of 20,000 or 50,000 or more, as a function of the characteristics of geography and demography, of sites and land use, of existence and proximity of services, of facilities, of touristic attraction, of local finance, and of human activities (such as factories), to be provided with ecological data.* This classification has been undertaken by the Ministry of Agriculture, using factor analysis.

This zoning, based upon multi-criteria analysis, would allow design of sample surveys in order to make the accounting, investigations and various reviews related to the environment, and also to ameliorate the network of observation of nuisances. Further, this zoning would allow one to go through and put together the various available information according to common framework adapted in a better way to environmental studies.

In particular, this classification would be used in the analysis of causes,

---

*To satisfy the same objectives one can also divide the national territory into identical squares, municipalities being then kept as elementary units.

effects and reactions of various factors that create pollution, nuisances, disfunctioning or amenities for human beings; in gathering the observation properly; even, if it is necessary, to finish off the available information by using more detailed analyses or surveys, for example: health problems analysis, social security data analysis.*

This zoning would also allow the study of problems very different, like migrations and population aggregates, or human activities, also problems like location of facilities and related costs in areas over- or under-populated.

In the same way, this zoning will be useful to set up and/or articulate a system of indicators relating to specific situations, emergencies, finance, management or even structures.

Land use statistics. It seems that special attention must be given to annual land use statistics because they are a preferential indication of the environment (concentration phenomena, desertification or migration). These concerns could be materialized in France by including an "improvement and future directions" section in the existing annual land use survey made by the Ministry of Agriculture.

It is an annual inventory made from human observations upon 720,000 places chosen at random. This sample survey allows the evaluation of changes in land use over the whole territory and projections over time owing to the permanence of the sample (Markov processes'). There are two directions to be followed by the proposed improvements:
1. Spreading out of this classification to non-agricultural land uses.
2. Trial for classifying, independently, observations according to a functional land use and landscape (see studies for the West Coast of the U. S. A.).

The obtained observations could be confronted with other sources—existing maps, cadastre, etc.—and eventually completed by surveys at a national scale like the present, and almost completed, one about forests. Also, one can mention the usefulness of remote sensing to observe, in detail, land uses and pollution of oceans, rivers, air and so forth, even noise of the traffic. Information is gathered by an airplane (satellite by 1980), then recorded and transmitted to the ground level, translated into numerical data and finally processed by computer.

Because trajectories of airplanes can be precisely calculated, it would be possible to proceed to different surveys made under identical conditions for the same area at different periods of time. It would, therefore, be possible to measure evolutions to make projections, and also to study several phenomena

---

*See "Case Studies on Middle-Size Cities."

simultaneously. (An attempt with measures on the ground at the same time is prepared for next June in the south of France.)

Direct pollution measures. The direct pollution measures must be technically and financially made by the services in charge of resource management or pollution control. They are dependent on engineers or technicians belonging to different public or private services. However, the statistician appears at the co-ordination level to make the choice of the various parameters and indicators to be observed for locations and extrapolation coefficients to be applied to measures (survey plans) and for the statistical criticism of homogeneity of readings for the aggregation of elementary observations.*

Measures of pollution—water pollution (rivers and oceans), air, noise, radioactivity, etc.—are still relatively scarce, and some important statistical problems remain to be resolved:

1. Location, frequency, and minimum number of data gathered in order to get a proper picture of the whole territory, divided into areas, with an acceptable precision level.

2. Minimum number of readings to be recorded, the precision of measures in order to easily observe variations over time, and minimum level of precision required given the technical possibilities for the experiment.

3. Weighing of measures for the same phenomenon in order to get significant indicators.

4. Selection of the most significant criteria and possibility to design composite indicators of heterogeneous phenomena that allow comparisons over space and time.

5. Coordination of locations of measurement systems for various pollutions in order to be able to study their effects in the same time.

It is necessary to foresee two series of measures in order to satisfy the different requirements of information:

1. A network of control, relatively light, compared to the number of stations and to the number of analyzed criteria; this network would allow a permanent observation.

2. Some measures with more parameters studied taking advantage of inventories of pollutions on a 3-5 year basis.

Noise is an important pollutant which sets its own difficulties. First, we note that physical measurements have not yet been developed and that we are only dealing with extremely narrow fields (mainly transportation).

Secondly, we cannot measure the state of the environment, only the nature and quantity of noise produced at a given time. Perturbations are

---

*Multi-criteria analysis.

dependent upon noise, and associations of produced noise, spectrum of acoustical frequencies, duration and continuation of noise, mobility of the source. Physical measures do not give full information about man's perception of phenomena whose psychological and biological equilibrium runs the risk of being affected. It would be suitable to gather some supplementary data consequent to physical measures, such as complaint lists, health statistics, opinion-perception surveys.

In France, the analysis of petitions mailed to the M.P.N.E. shows that people are especially sensitive to pollution exacerbating the senses: air pollution, noise, odors, and aesthetics, which are hardly quantifiable with physical measures.

Statistical information provided on wild life, especially fresh water fishes and game, are very fragmentary for the peopling and the stocking, as well as for the breeding, the production, and the mortality due to pollution.

It seems to be useful to gather new data on animal life, especially on fresh water fishes and game for the reason that:

1. Data gathering used as an indirect indicator to observe the state of the environment with respect to the pollutants which would take place in the eco-system description.

2. Useful information gathering in order to set ecological or patrimonial accounts as also a nature audit.

Managerial and structural or conjunctural indicators of environment. We can propose several kinds of indicators:

1. Conjunctural, or, for alarm, permanent, relatively simple, and if we are willing to prevent pollutions over critical levels (in France, for instance, pollution control of air in the Rouen Valley).

2. Short-range management (one year for instance), if we are willing to focus on some aspects of the state of environment and especially of its evolution in order to act within a budgetary and a statutory framework; for instance, quantity of pollution discharged into continental waters or polluting collectivity tax basis; used series are more numerous but observed with less frequency than previously.

3. Structural, when we are willing to put in relation levels or to establish functional connections and behavioral models; these studies are complex and lead to an accurate and spaced information gathering; for instance, quinquenal inventory of water pollution.

The last analysis group requires some critical and statistical treatment works, sophisticated enough to choose with judgment the resumed indicators used in the first two points of view; for instance, minimum level of parameters to gather at a given point in order to characterize the water pollution.

In this field, we can notice a multi-criterion analysis research (factor analysis of correspondence) on hand in France and dealing with about 60 variables monthly, gathered over a year; the topic is rivers and canals water quality over 100 stations, with flow and bacteriological and physiochemical analyses. The subject is to obtain the weight of each factor, and, if the levels of the results will allow it, to limit the more determinative components by balancing them with regard to their relative value in order to obtain synthetic indexes.

We also will notice that very often indicators are dependent upon the state of technological knowledge (for instance, presently in France we can't study the white dust spread in the air), and that very often it is not sufficient to observe standards overstepping but also to know the intensity and the duration of the crucial phenomenons in order to evaluate their effects. Moreover, standards can be different over time, between sectors, regions and countries.

So we have to gather data over the totality of the phenomenons distribution and not only over some characteristics of this one.

Finally, we will notice that most of the time, proposed indicators are only pinpoint statistical series or ratios. Substantial findings are necessary in order to obtain synthetic indicators (research of an indicator characterizing water pollution, for instance), to aggregate observations (proper balance of observations on air pollution by residents and workers within concerned area, for example), and to establish a coherent and adequate indicator system.*

Establishment of stocks accountings, materials audits and ecological accountings, improvement of national accounting with regard to environmental concerns. As data gathered about the environment are available, we have to think about taking them into account at the level of global and synthetic statistics.

The interest of establishing patrimony accountings (lands and forests, natural resources, buildings, wildlife) is great because they would lead to a descriptive audit of the nature and to balance on it flows and consummations with existing stocks.

It also could be interesting to work out materials accountings for production inputs and consummation outputs, recycling or dispersed waste over the components of the natural environment.

These input-output tables would be very useful for the analysis of

---

*For all metropolitan areas, we can design a systematic network of indicators perfectly homogeneous and convenient, conjunctural, managerial and structural as well. However, particularities of metropolitan areas (climatology, site, hydrology, communication network, location of facilities, of industries, of residences, etc.) make it necessary that a series of supplementary indicators related to each center will have to be added to a national or international list.

pollutant flows on all levels of production, processing and utilization as well as the physical inputs and outputs not noticed on the market, and divided into polluting and non-polluting products. We will establish material accountings into physical units, and, if possible, into value. These works would help to set up an evaluation of deterioration of the environment, if we connect them with stocks study and with the programming. The previous two sets of works will permit one to obtain ecological accounting describing the mechanisms and relationships between the natural and artificial eco-systems of utilization and transformation of existing elements according to adjusted classifications.

Besides these new studies, in order to include environmental quality preoccupations into the national audit system—without any changes of its basic framework which has its own utility—it is possible to adapt existing audits, for certain specific studies by inclusion or modification of certain classifications and concepts and by the introduction of the agent "nature".

The development of the studies on "material balance" are linked to the making of a specific inquiry by highly specialized inquirers into the so-called polluting industrial plants. One would gather in these plants, data about the technological processes of production, the flows and functions of pollution, the material inputs and outputs, and the costs and investments in pollution control.

These studies on material balance are linked to the existence of data on products which might have toxicological effects: production, transportation, consumption, diffusion into the environment which could initiate the definition of a sensible middle- and long-term policy about the use and treatment of those products which are to be classified according to appropriate classification.

Case studies on middle-size cities. In France in the frame of the Inter-Ministerial group of Evaluation of the Environment, it has been decided to study one city and its suburb, other than Paris and the regional metropolitan area. Rouen has been chosen.

The purpose of the study is to make a methodological and systematic pre-census of the various environmental problems and of the various elements which contribute to the quality of life.

This should enable us to:

1. Establish a method of approach to complex problems and to the study of the quality of life in an urban setting.

2. Build indicators of the quality of life which would take into account the interrelations of the problems.

3. Look for the articulations and interactions between the elements and man in quite a detailed way, in order to make explicit the processes to prepare decisions (see zoning).

It is thought to analyze, during the 1974-80 period, several cities of 50,000 and over—one a year for instance. Each of the studied cities will be considered as representative of situations and problems existing in other similar urban centers.

To end this talk, I would like to say a few words about the factor analysis of correspondence which is a typical French methodology. This method has been discovered by Dr. Benzecri of the University of Sciences of Paris and he is also a Dr. of Princeton University. The method was discovered eight years ago. This method of data analysis is objective, and the cost in computer time is not consequent. The method has been published in two books and it is possible to explore a great number of different data to classify and cluster the units and the characters by a measure of proximities, extraction of factorial axis, projection of units and characters simultaneously, explanation of axis, which are in combinations of scattering, to better reduce with minimum loss of information and the characters appeared in order of importance.

DISCUSSION

OPPENHEIMER—This paper is now open for discussion.

BROGDEN—How many synthetic indexes are you trying to produce? For water quality would you just try to produce a single synthetic index or would there be more than one?

LENCO—One only.

BROGDEN—But how many would you aim for, one or two?

LENCO—Two or three-part indexes. It's just beginning in France now and the main problem is to gather and to class the information in order, but it is not the case at all, the Engineers and the Chiefs of Information Centers are links and they have not a care of uniformity.

OPPENHEIMER—Do you have one data system in France for all of the environmental information?

LENCO—Oh, yes, a scheme.

OPPENHEIMER—One system, and it will be a government operated system?

LENCO—For water, it is a specific organization which is national, but there will be information in private sources, or industry, or towns, or state organizations.

OPPENHEIMER—Your environmental data base is nationalized?

LENCO—Yes, ten years ago.

NELSON—Is this self-supporting system the answer then, that is my question. You said it is regulatory or is a self-reporting system by the industry or the town; they report to the government on a self-reporting basis? The government doesn't go out and take all the samples and put it in the system? They take the information from the city or the industrial component and compile the information from them, is that the flow of data?

OPPENHEIMER—Does the industry give the information to you?

LENCO—Yes. We have general publications also.

BERG—Is industry obligated to collect this information through legislative means, or is this accomplished otherwise?

LENCO—That's the special device of the Minister for the Environment in France, he is charged to collect information and to implant the stations, for example. But for industrial it is very particular. For the small industry there is no difficulty, but for other factories it is not easy. We have difficulty because the taxes are published, and calculated with productions, and they are obtained by the boards and staffs that supply information.

LIGHTHART—This is sort of a nasty question but I just have to ask it, how do you know how reliable the information is that comes from the industries?

LENCO—Because it is necessary to make statistical inquiry by sampling.

LIGHTHART—What if the industry continually lies to you? Then you get a nice small variance; you get a good statistic but it isn't the truth. Then what do you do?

NELSON—A case in point is to assess A, B, and C polluters. We approach the environment from many data packages. When you are looking at point sources, you can trace the degradation products, and after a period of time isolate the source of contamination. In our sampling, we hire consultants, take industries' and municipalities' data, and collect data under cooperative agreements with the U. S. Geological Survey.

BROGDEN—Didn't the Environmental Protection Agency get involved with coming up with a single number to represent air quality? How did they approach this?

HAMMERLE—Well, that concept has never been accepted and is a very complicated scheme. As you can imagine everybody had their own way of evaluating air pollution. But you are correct, that is not part of computerization. Really, that's an air pollution information problem and has nothing to do with data-backed operations because the problems are always incognizant of the engineering things that you take in order to come up with this index. It has not been acceptable so far and I don't think it will be in the near future.

Of course, ultimately we will probably be able to come up with something but it is not possible now because we have a tough problem. We have fixed monitoring sites for air pollution and you don't know how to take the daily fluctuation on one side and combine it with the daily fluctuation on another side and come up with a measurement for the air quality in a county, let's say, in which we may have two or more data collection sites. How do you express it? You take a reading every minute, and it changes throughout the day, and you try to come up with something that represents that county, let alone something that represents all the air pollution for some geographical area. It's extremely difficult.

And the other question is how good is the data from the source. I have as my responsibility a source inventory of 75,000 point sources of air pollution in the United States and the problem we are confronted with all the time is just how good the data are. And it varies all over the map from such instances as a large steel company in Pittsburg saying we didn't know that we were supposed to report anything, with a hundred thousand tons of sulfur oxide emission; to a situation in Florida, for example, in which every single source of air pollution is reported. One tends to have pretty good confidence in the data from Florida. But in Pennsylvania we have a situation where you know that most of the polluters haven't even filed, so it's difficult to have confidence in the reported data. The only solution is to actually test the source of pollution yourself, and that is extremely expensive. It varies anywhere from 5-50,000 dollars per single test at a pollution source, and there are 75,000 sources in the United States with over 100 tons per year. With an average of $20,000 per source tested, you can see the magnitude doesn't immediately compute for guaranteeing to have good quality data. This is where the cost effective approach comes in.

KITSCHLER—May I remind you that the statistician normally does not venture an opinion. By some, a statistician is seen as a "back room boy." I should like to comment on Dr. Hammerle's remarks, especially on the problem of computing a correct gross national product. We face a situation in which all the figures used just aren't accurate because of the basic figures that we use for water and water quality, but especially the figure that we use for air. Air is considered a free product which can be used just as anybody pleases. And my view is that proper statistics can only be achieved when it is possible to make a liaison between the big political issues and the statistical methods to show them in figures, maps, etc. Are there any studies in this field? In France especially?

LENCO—In France we have a great deal of difficulty with them, the amount is plenty.

HELMS—I would like clarification of Dr. Kitschler's point of view. Do you suggest that certain natural resources like water are included in the gross or the net national product? Was that the essence of your statement?

KITSCHLER—It was.

HELMS—I can only say that it does seem to me that these mappers that are economists should learn that aspect as soon as possible. Because in a way the pictures we have available now, particularly if we look at things on a long-term basis, they are just teasers.

OPPENHEIMER—Any other comments? I should think that this aspect of statistics would enlist quite a bit of response by many people. I do feel like Dr. Kitschler that statisticians for the sake of mathematics may have distorted many things in the past.

BROGDEN—I know that in some cases the air pollution people, if I remember correctly, used a geometric mean rather than arithmetic, and there is a certain rationale behind that. What is the feeling; do you use this for every type of summary?

HAMMERLE—That is a standard, as a matter of fact, for air quality measurements. We do both the air at a mean and standard deviation and geometric mean and standard deviation and frequency distribution, so different applications require one over the other. Our data facts report all of this; it has a standard outflow and with it our problem is that very seldom do we have anyone interested in our raw data file. People are always after our summary statistics simply because the raw data files are too large for anyone to work with. So everything is always reduced to summary statistics for a specific pollutant or a specific monitoring site because we don't know how to combine two sites and we don't know how to combine two pollutants. We don't even know how to combine two methods of measuring a single pollutant at a single monitoring site. So we have to work with summary statistics.

BROGDEN—Do you ever go back to the raw data or are they just not available?

HAMMERLE—We keep it, but for air pollution work people normally operate on the summary statistics and not the raw data. Our monitoring sites are fixed and we have somewhere around 6,000 monitoring sites in the United States, and most of them monitor five pollutants on one of those continuous devices that take a reading every three minutes. So the numbers explode pretty quickly.

OPPENHEIMER—I would like to ask one question, is it France's intention to make its data system available to the public or will it be proprietary?

LENCO—Yes, it's the first hypothesis. In principle, it will be open to the public, perhaps with exceptions such as health problems and scarce toxic compounds.

OPPENHEIMER—Will all requests be honored?

LENCO—Right, but now we don't know exactly. The system will be useful for studies.

OPPENHEIMER—I see, you are just developing it.

LENCO—Yes.

KOVACS—Dr. Hammerle, in the air data, can you specify or break down summary data by industry, or is it just by location or pollutant only?

HAMMERLE—You are talking about emissions?

KOVACS—Yes.

HAMMERLE—Not air quality data, emissions? We can retrieve for any geographical area down to a county for any industrial category. And we have more than the standard industrial classifications, there are 900 source categories. We have developed our own, it's a finer gradation than FIC's. So the answer is yes, by pollutant, by industrial classification, by geographical area, or by individual source.

Now the problem is that in air pollution, emissions are calculated and you calculate them using what's called an emission factor, that's a number that you can apply to what is done at that facility. In order to calculate the emissions, for example, the emission factor might say 50 tons of particulate are generated when you perform a certain operation at a certain rate so we have to know what that rate is in order to apply our emission factors.

The problem lies then in the activity that is taking place at the facility. If it is patented or has proprietary ownership on the operation, or if that number is directly associated with production, that particular facility does not want that number made known to the public, even though that's the number that you have to apply your emission factor to. For example, so many tons of soap are manufactured per year at a soap factory. The factory does not want anyone to know how many tons of soap are manufactured or how many tons of a certain chemical are manufactured because they say that is proprietary information that gives everybody an idea of exactly what that business is doing. And in that instance, you may have the same situation for water pollution and so on, because if you just say I know that soap manufacturing or dye manufacturing requires a certain amount of water in order to produce a certain amount, then you can just multiply backwards—take the pollution, divide by the emission factor and you know what their production is. These plants do not want that information released.

In fact, in Texas, one of our biggest problems with respect to proprietary information is that the State has required that everything, including name and address of the facility is to be kept confidential. More than half of our confidential data pertains to Texas facilities, and this is really a problem.

PEACHEY—Mr. Chairman, I think we have hit a rather delicate but quite important stage in our discussion. I think that the over-valuing of the commercial significance of information is a form of conceit, particularly applicable to the developed world and that as we advance in data management and data pooling we shall have to give something to people in exchange for what we take. What we really give and take must be in the context of a new kind of democracy of information in which people are able to check what is said about them and their activities.

We have not, as far as I know, as a race, consented yet to the acquisition of full data about ourselves. These data are probably the best guide as to what is helping us, or poisoning us. Our own human tissues as an indicator can do a considerable amount of internal arithmetic on the balance of pollutants and the way in which they reach the human sink. My own purely personal feeling is that more epidemiological statistical work needs to be done on man. It's far from easy to measure the exposure of the human population. It's much easier to measure the input of pollutants into various media.

I think that in those countries that have national health services, sooner or later, we are going to get nationwide patient records banking, again with appropriate safeguards, but which for the first time could enable us to achieve new epidemiological and historical correlations. I think that when we see these patterns of disease and mortality, we shall begin to get a picture of human health and environmental health which will totally transform our need to monitor in other ways, and that until we are brave enough to take that step with all its inherent problems, we shan't really significantly move forward.

OPPENHEIMER—I agree, there is a very good example of why we do need data to try to combat some of the current misuses of environmental control. DDT control is a good example. Before Rachel Carson brought attention to DDT, we had malaria pretty well under control. Seven years afterward, with all of the environmental constraints on the use of DDT and the worldwide governmental restriction of DDT material, the people in Ceylon have over a million cases of malaria. In our own country, Miami, Florida, was very badly hit with the ban on DDT and its use has been resumed by special case.

Here is a very clear example of where data can help balance the environmental constraints that are being established. One of the main purposes of this Conference is to try to formulate ways in which we can manage data so that valid action results in the future.

FLEMING—EPA is monitoring pesticides, using us for analyses, I think its fourteen or fifteen major sites in which they are making analyses on the effects of pesticides on people and they are using statistical analyses.

EBERHART—I just wanted to mention one other study. I don't know if it is published yet, but EPA has done a study called the CHESS study on human effects of sulfur dioxide. It is some 800 pages long, and a very thorough analysis of the effects on human health of sulfur dioxide; but, this kind of thing will have to be done with greater frequency and for many other pollutants in the future.

I just wanted to mention one other thing. It seems to be a very difficult thing to separate effects of different pollutants. If you are in a metropolitan area, and you have an increase in some kind of disease or other, it is very hard to tell whether that came from sulfur dioxide, or nitrogen oxides, or hydrocarbons, or particulates, or what. And I think this is an area that deserves much more study.

OPPENHEIMER—I believe this is a good point at which to close this session. I would like to thank Mr. Lenco for his stimulating talk and to say how pleased I am at the free exchange of philosophy and information that has taken place during this Conference so far.

VI. THEORETICAL CONSIDERATIONS

Session Chairman
Dr. Martin N. Cobb
Environment Canada
Canada

Principal Speaker
Mr. Henry Fleming
Gulf Universities Research
Consortium, U. S. A.

COBB—The subject this afternoon is "Theoretical Considerations," and Mr. Fleming's talk will be specifically related to the ENVIR System which has been developed by the GURC (Gulf Universities Research Consortium) group, Mr. Fleming.

"A Computerized System for Handling Environmental Data (Analysis, Synthesis and Display using Logical/Scientific Environmental Principles)"

Mr. Henry Fleming

The EDMPAS System (Environment Dependent Management Process Automation and Simulation) consists of an analytical retrieval program, ENVIR (ENVironment Information Retrieval), that enables the user to partition the data bank into the required subsets and to select the necessary data from these subsets to be input into a series of modules (Figure 36). These modules consist of X-Y Plots, map plots, simulation and calculating programs, clustering programs, information analysis programs and ecological programs.

The characteristics of ENVIR are in brief:

1. Free field input.

2. An input "skip and dupe" facility to simplify the input of redundant or missing data.

3. The ability to address any bit (piece of data) by Boolean expressions.

4. Arrange any extracted information hierarchically.

5. Arrange the output format as desired.

6. Order the information as desired.

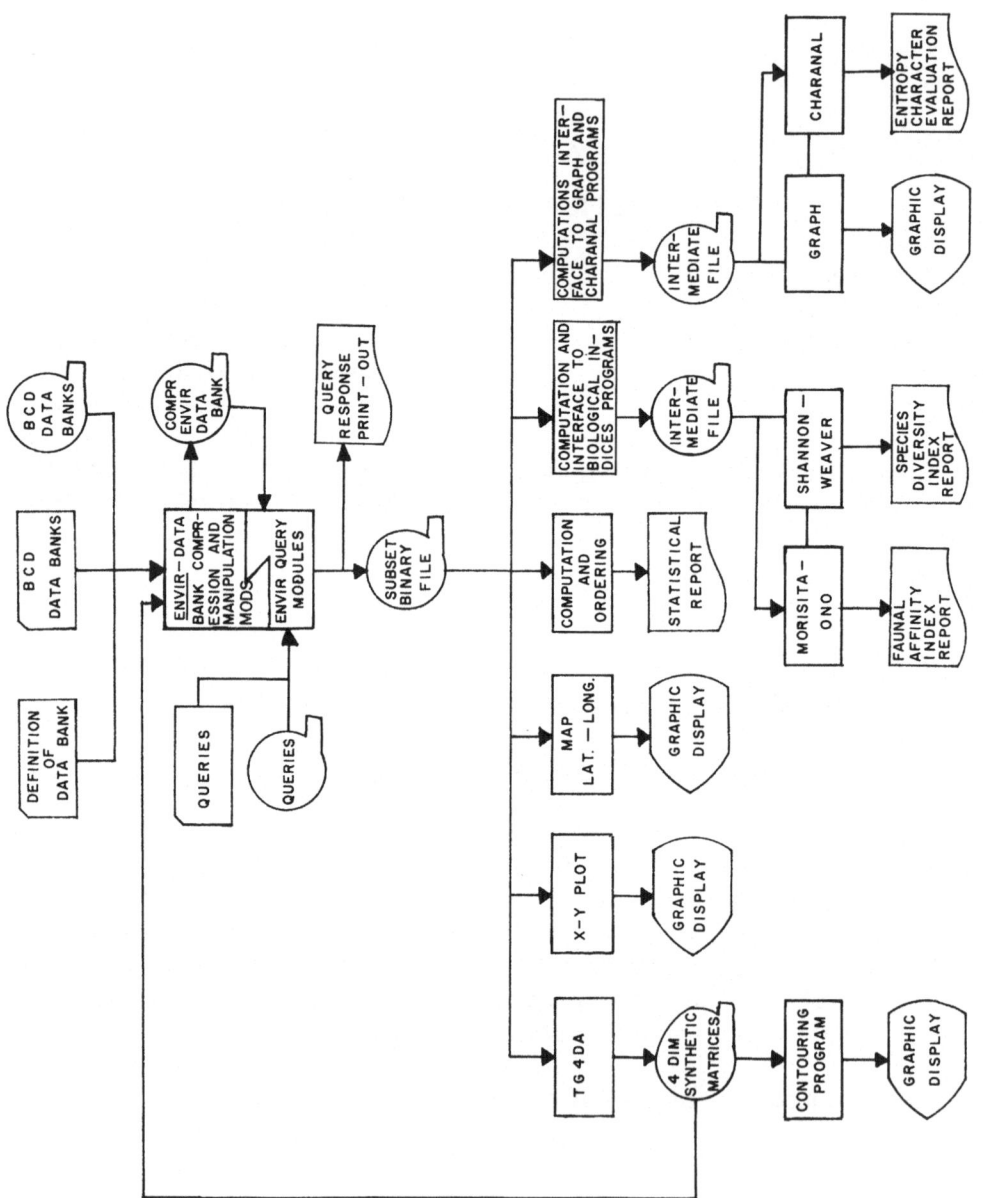

FIGURE 36. EDMPAS system development status.

7. Output the data to selected peripheral equipment (hard copy, CRT, plotters, etc.)

8. Correction information.

9. Delete information.

10. Add records.

11. Add fields.

12. Rearrange the structure of the data.

13. Redefine any data fields in core.

14. Create new fields by the manipulation of existent fields.

15. Compresses most data banks by at least 1 order of magnitude (Dynamic Information File).

16. Adapt program dimensions to computer core size or cost benefit considerations.

ENVIR possesses a number of other capabilities but the above listing is adequate for the majority of environmental cases.

Since the purpose of this discussion is to show the applications of information analytical methods to the resolution of environmental problems, the remainder of the dissertation will briefly explain and illustrate the applications of some of the ENVIR modules to environmental cases.

TG4DA (TOPOGRAPHICALLY GUIDED 4-DIMENSIONAL ANALYSIS). This program has application where environmental conditions are physical and these fields are space-time dependent; in other words, when the physical parameters can be thought of as continuous even though variable. For instance, given a number of water/air samples of concentrations of some pollutant collected at a few known points of space (Latitude, Longitude, Depth) and time, a much denser field of these values may be created that can be over-laid and correspond with the field desired from another series of biological samples taken in the same area but with different spatial and temporal points. Thus, variations in species density can be detected with changes in concentrations of chemicals or other physical parameters. In practice, it has been customary to illustrate the correspondence by using the results of the analysis in the MAP-PLOT module of EDMPAS. The axis would be some range of Latitude-Longitude selected by the investigator and a fine contour line selected to represent some physical or chemical parameter and a dotted line to represent the species frequency contour.

Principles of operation. The environmental quantities to be measured—temperature, salinity, fluid velocity, and others that can be derived from them—are physical fields continuous in time and space. Short of representing them by analyzed functions of x, y, z, and t which we do not possess, the closest similitude, and the one most amenable to further computer processing, such as by means of models, is to store them in the form of arrays of values of each field at every point of a grid covering the domain in which observations were made—in the case at hand, the Cayman Trough.

The process by which this array of grid-point values is obtained from the observations is known as OBJECTIVE ANALYSIS in meteorological jargon and is commonly executed in two-dimensional form, resulting in a map of an atmospheric field at a certain time and altitude based on synoptic observations. However, a technique of objective analysis has been developed by which the fields are defined and stored as four-dimensional arrays with two very important advantages—that the inherent space-time continuity of the fields is embodied in the mathematical definition of the output arrays, and that the program interpolates in time and height (or depth) as well as in the other dimensions, and is, therefore, no longer limited to the use of observations taken at standard times or levels, but employs all data gathered anywhere in the x, y, z, t domain explored.

The method was derived from the familiar "first-guess-and-correction" scheme proposed in the fifties by Panofsky, 1949; Cressman, 1959; Bergthorsson and Doos, 1955; and others. However, the first guess here is the field's coarse structure as defined by a first look at a sample of the data, whence the name: Topographically-Guided 4-Dimensional Analysis, condensed in the acronym TG4DA. Thus, the process is entirely data-based and no longer needs an initial assumption of the grid-point values.

The field's structure is defined as follows. The program retrieves the observations pertaining to one or several levels specified by the user. At each of these levels, it determines the mean value of the corresponding data and from it, the areas where the field is high and low. Highs and lows are extended into all dimensions to fill the array and in each of these partial hypervolumes a quadratic function of all coordinates is defined by a least-squares fit to the data contained in the hypervolume. The resulting hypersurfaces are then meshed together in the evaluation of a complete set of grid-point values, which forms the First Approximation Field.

This First Approximation Field is then iteratively adjusted to the observations by a four-dimensional, nondisciplinary algorithm. Each grid-point value is corrected to minimize the errors at the surrounding data points contained within a fixed radius of influence. One such radius of influence is defined along each coordinate. The user specifies its initial value. It is then halved for each successive iteration in the phase.

The rms difference between the observations and the analyzed field is computed at the end of every iteration. When it reaches a target figure specified by the user, or when it fails to decrease at a rate that justifies an additional iteration, the analysis is terminated.

The program is written in FORTRAN V and runs, at present, on UNIVAC 1106 and 1108 computers. Its plotting segment was designed for the SC-4020 CRT display unit, but contains a library of interfacing subprograms that enables it to accept CALCOMP commands and, therefore, makes it operable on any CALCOMP-compatible plotter. This segment drives an isopleth generator developed by the NOAA National Hurricane Center, program ECHKON, which provides contoured cross-sections of analyzed fields. The input consists of the user's data, on cards or BCD tape, and a set of control cards whose contents are used to set up two Control Blocks that direct the entire action.

The Control Blocks contain, among other things, the parameters required to define the grid for the analysis. On this bases, input data are first converted into a time-sorted binary file in grid coordinates, the DATA file, which can be fed-in directly in later runs. The distribution of data along the vertical and time coordinates can be printed out to assist the user to set up his analysis.

The analysis itself is an overlay segment that uses the DATA file and creates from it a two, three or four-dimensional analyzed field, depending on the request entered in the control cards. This field is stored in a file with the contents of the Control Blocks, and can be saved on tape. This makes it unnecessary to reanalyze the same data when more plots of the field are desired. In such cases, the Control Blocks of all analyzed fields in the file (up to six) can be printed out to assist the user in setting up his plots.

Figures 37-40 represent various types of output available from the TG4DA module.

Figure 37 illustrates the temperature variation in a selected cross section area from a much larger data bank. The small figures at the extreme left are the depth in meters and the large figures at the left and right are the temperature at the respective contour. The minus signs within the body of the figure indicate the points within the field at which data had been collected and input to ENVIR and the plus signs are the points for which TG4DA computed the values.

Figures 38 and 39 are XY Plot and Map Plot Modules. At this writing, the X-Y Plot and Map Modules have been combined and generalized. The X-Y Plot Module has been used not only for illustrating output from the ENVIR statistical programs and other data field X-Y correlations, but also for mapping collection sites, etc., where the area covered has been so small as to make map projections inconsequential. The symbol and contouring facility has been retained and the various map projections such as UTM and mercator are at the users discretion. Computer time for Figures 38 and 39 was 4 seconds each.

Figure 39 is a sound velocity profile after the requisite data was selectively retrieved by ENVIR, processed by TG4DA, the sound speed calculated by Wilson's equation from the TG4DA output and then the results plotted by the ENVIR X-Y Plot Module. Sound velocity data of any cast at a specified location and time can be selectively retrieved from ENVIR dynamic file and displayed in the form of sound velocity profiles.

Figure 40 illustrates the tracts of four trawls taken to collect ground truth at approximately 7500 foot depth. At the same site several thousand photographs were taken and the benthic organisms identified by use of the organisms collected with the trawls. Sediment characteristics, chemical and physical parameters current direction, etc., were also collected and incorporated in the ENVIR DYNAMIC FILE. Rather than attempt to include an excessive amount of information in any one plot such as species distribution, percentage kill contours, sediment characteristics, etc., a number of different plots were made and developed into transparencies so that the various plots could be superimposed

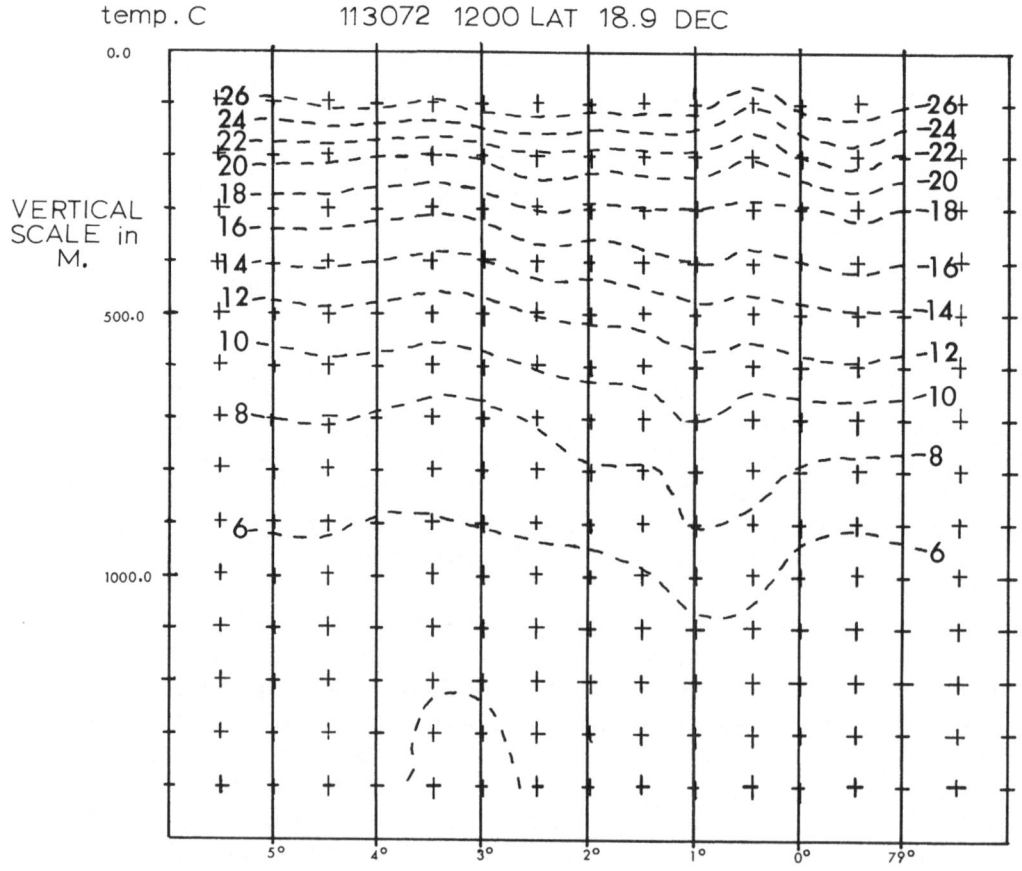

FIGURE 37.   TG4DA zonal section.

on conventional maps since scale factors, map projections and azimuth and range requirements and characteristics can be entered in the program.

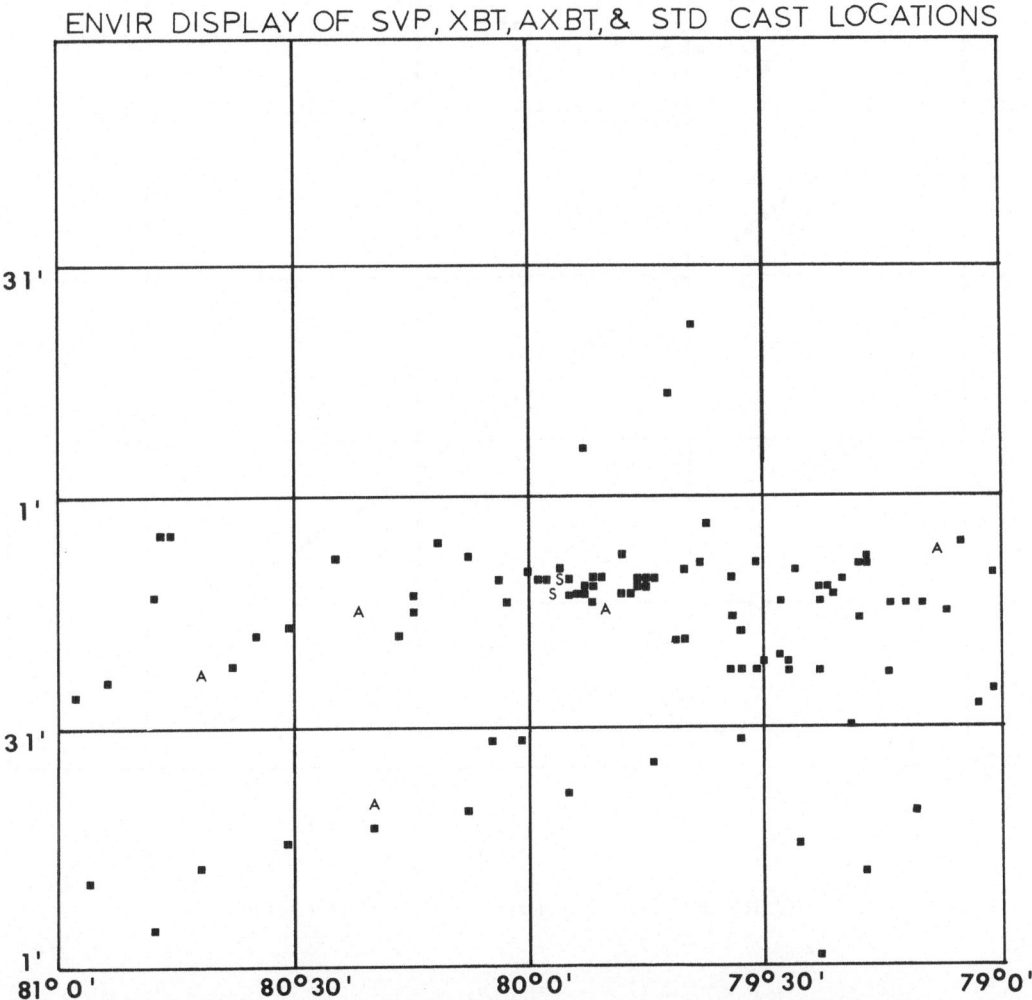

FIGURE 38. ENVIR map-plot of the locations of SVP, XBT, AXBT and STD casts.
S = XVP; ■ = XBT; A = AXBT; T = STD

GRAPH. The primary thrust of this module is the clustering of qualitative data, characteristic of much of the type of information contained in biological, ecological, medical, etc., banks that are not adaptable to the usual statistical routines. It is also based, as noted earlier, on the premises of <u>overall similarity</u> rather than a selection of an ordered range of parameters.

The program permits the definition of three different types of parameters or descriptor fields:

depth in meters

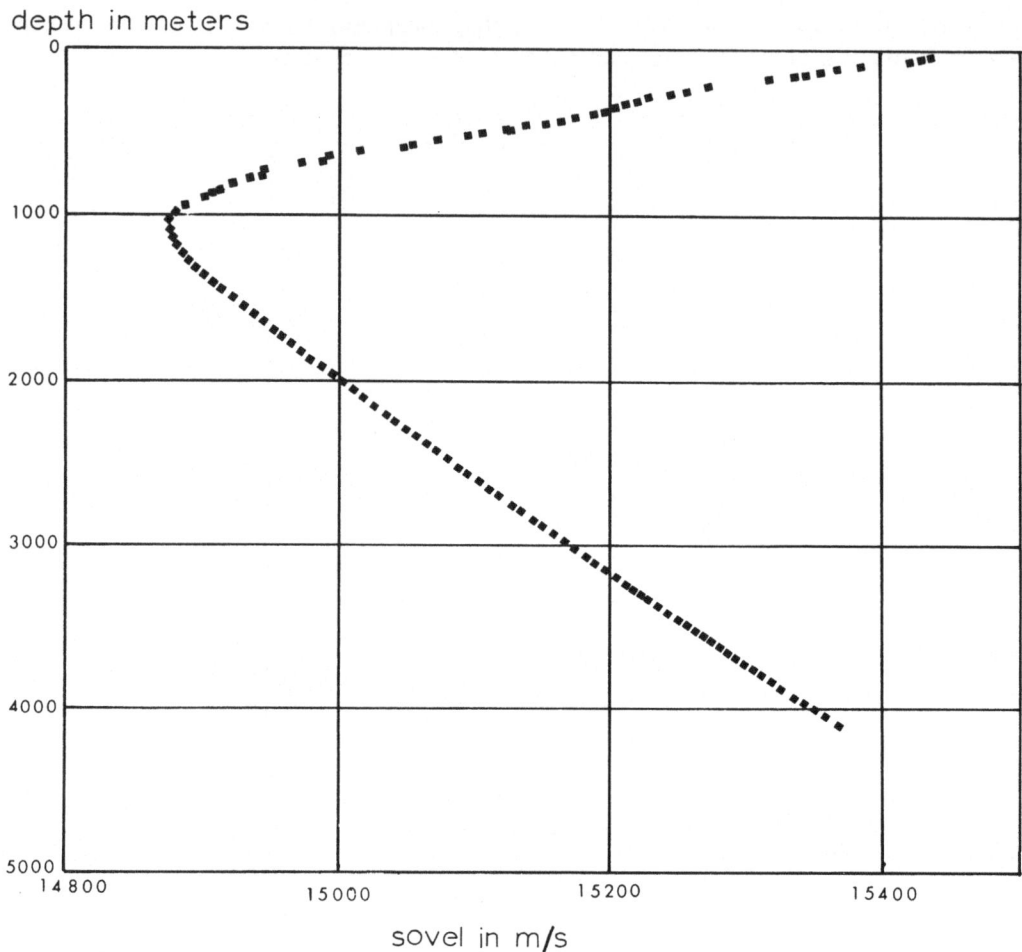

FIGURE 39. ENVIR generated sound velocity profile.

1. Those descriptors in which the recorded state must be identical to be considered as a match.

2. Those descriptors whose states have some regular order such as very small to very large and it is desirable to indicate a similarity between objects over a range of states. Those may, of course, be a set of absolute numerical values. The following equation is used by the program to calculate the similarity for this type of descriptor:

$$S_{ij} = \frac{2(K + 1 - d)}{2K + 2 + dK}$$

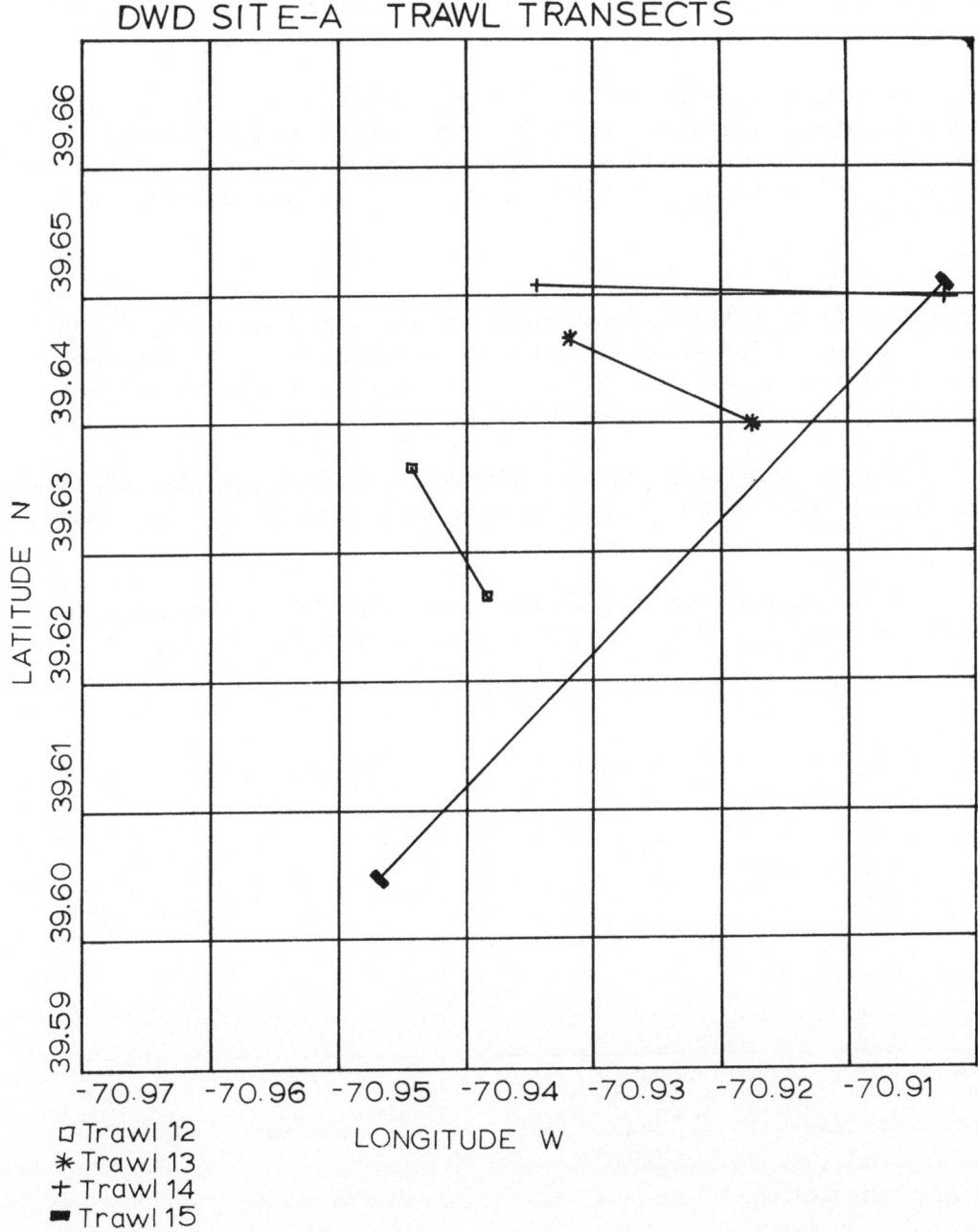

FIGURE 40. Trawl transects showing the tract along which biological
data were gathered. (An example of EDMPAS map-plot
module output.)

where Sij is the similarity of the objects i and j, and d is the distance between the states, K.

A third type of descriptor frequently occurs in environmental data in which the states of a descriptor show a nonidentical relationship that is not well-ordered. Under such conditions, the worker is allowed to input his judgment of the degree of similarity to be reflected between the different states assumed by the objects.

Provision is also made for not having certain descriptors that are inapplicable for a similarity comparison because of a logical default of the data to be entered in the calculation of similarity. For instance, in a data bank of organisms containing mature and immature individuals, sexual descriptors must be restricted to a comparison between mature individuals.

Also, when some objects lack information for a descriptor(s), this/these descriptor(s) is/are excluded for the respective object(s) in the similarity computation.

Procedure and printout. (See Figure 41 for printout. Alphabetical letters in brackets in the text refer to respective sections of this figure.)

FIRST PARTITION: L = 1, C-value = 0.9500

FIRST CLUSTER MEMBERSHIP	[A] C-value	[B] CONNECTEDNESS		R (1)		
1, 2	.9500	1	1	(1,2)		
MOAT = .1132	NEXT PAIRS TO JOIN (2,6)					
SECOND CLUSTER MEMBERSHIP	C-value	CONNECTEDNESS		R (1)		
3, 5, 7, 9	.9500	3	6	(3,5)	(5,7)	(7,9)
[C] MOAT = .0500	NEXT PAIRS TO JOIN (3,4) (7,8)					
SINGLE MEMBER CLUSTERS (4)						
4, 6, 8, 10						

SECOND PARTITION: L = 2, C-value = 0.9000

FIRST CLUSTER MEMBERSHIP	C-value	CONNECTEDNESS		R (2)	
1, 2	.9000	1	1	--	
SECOND CLUSTER MEMBERSHIP	C-value	CONNECTEDNESS		R (2)	
3, 4, 5, 6, 7, 8, 9	.9000	8	15	(3,4)	(7,8)
INTERNAL CONNECTIONS AT .9000		INTERNAL CONNECTIONS AFTER .9000			
(3,7) (3,9) (5,9)		--			
MOAT = .0111	NEXT PAIRS TO JOIN (9,10)				
SINGLE MEMBER CLUSTERS (2)					
6, 10					

FIGURE 41. Printout of statistical program.

After reading the input data and checking its validity, the program will print out a copy of the input data at the users option. The program will also exclude any identical objects and print out a list of any duplicated objects.

Each object is compared with every other object for the purpose of computing a similarity ratio expressing the degree of similarity between each object pair. The similarity consists of the sum of the number of matches between the states of each descriptor, or the degree of similarity assigned to each state by the above equation or by the investigator, divided by the number of actual comparisons made (missing states or inapplicable states are not counted). The similarity ratios with associated pairs may be printed out if wanted.

A partition (L) of the study will occur whenever, in this ranked list of similarity values, a value changes. Consequently, as the list of similarities are scanned, whenever a lower similarity value (C-value) is defined. A cluster at a partition is a set of objects that are interconnected at any given level of similarity. In other words, there is at least one continuous pathway of connections joining all the objects. It follows that there may be 1 to N number of clusters at any partition and that any object entering only 1 of these clusters at a lower similarity value will create a new partition of the study at the latter value. The second cluster of the first partition L=1 may be represented as:

3-5-7-9

and at L=2 the further development of this cluster as:

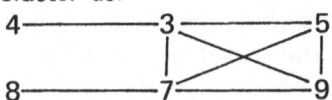

We term those connections that form new clusters (any object entering a cluster forms a new cluster considering that each object in the study before synthesis is a single member cluster) R connections. Any connections between objects within clusters between partitions are internal connections. The degree of internal connectedness is a measure of homogeneity of the respective cluster. The maximum connectedness is determined by

$$\frac{N\ (N-1)}{2}$$

where N = the number of objects in the cluster. In the printout at [B] two figures are given. The first figure is the number of CONNECTEDNESS between the objects at the respective level L and corresponding C-value or similarity value. The second figure is the maximum number of possible connections.

The MOAT [C] is a measure of the isolation or disfunction of a cluster. It is equivalent to the degree of relaxation of similarity necessary before the respective cluster will be changed by the addition of another object(s). The last partition will be that point at which a similarity value has been reached that allows all the objects in the study to be contained in a single cluster.

The C-value at [A] is the similarity value at which the respective cluster was formed. NEXT PAIRS TO JOIN permits the investigator to look ahead and know which object in the study will change the composition of the respective cluster.

Two additional printout options are available:

1. A model distance array which is a list of the 10 most similar objects to each object.

2. A histogram (Figure 42) showing the clusters and the disjunctions between them.

Obviously, the kind of objects clustered may be specimens, species, people, environments, etc. The restriction is whether or not the principle of "overall similarity" applies.

CHARANAL (Character Analysis). Two broad types of information are usually contained in an environmental data bank. The first kind of information is of an inventory or catalog sort of data. This data is seldom of interest for purposes of analysis or synthesis since it is essentially "identification" data. The second kind of data is scientific, in the sense that it can be meaningfully used for analysis and synthesis. This is the sort of data in which dependencies and correlations are of interest.

The answer to the following questions in regard to the latter type of data are of primary concern to any environmental investigator:

1. What is the classification significant information content of each character?

2. What is the amount of correlation between characters?

3. Is there any redundancy because the same information has been unintentionally described in different ways?

4. Should the descriptor states be redefined?

5. What is the diagnostic value of a descriptor?

6. Which descriptors contain random information or at least do not contribute to the purpose of the study?

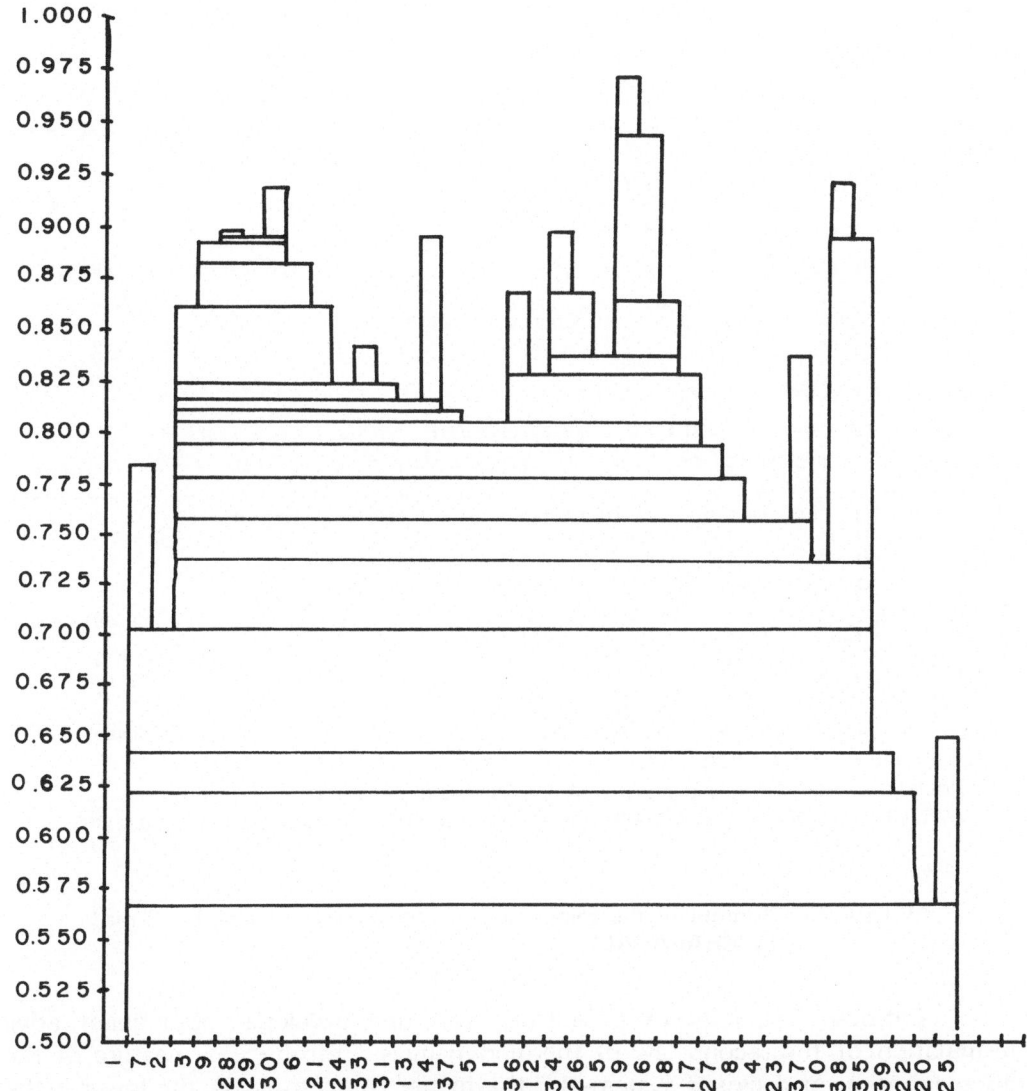

FIGURE 42. Histogram printout of cluster analysis.

If the descriptors in a data bank are composed of states that are orderable (the states can be associated with real numbers or placed in order on an axis) the usual parametric statistical methods can be used. However, many, and in biology, most, descriptors are of a non-orderable type. CHARANAL was developed to handle the latter type of descriptors, though it can also deal with ordered descriptions. CHARANAL used the actual distributions of the data rather than abstractions derived from the distributions such as means or standard derivations. It uses information theory rather than the theory of probabilities (see Figure 43).

[A] I = 1, J = 2

[B] CHARACTER I = 1 (6 STATES) COMPARED WITH CHARACTER J = 2 (5 STATES)

[C] D (1,2) = .67554   [D] S (1,2) = .32466

[E] OBJECTS NOT IN THE COMPARISON = NONE   [F] PROBABILITY OF GOOD OBJECTS = 1.00000

[G] PROBABILITIES OF THE STATES OF CHARACTER 1, GIVEN THE GOOD OBJECTS
```
       I(1)     I(2)     I(3)     I(4)     I(5)
      .10000   .43533   .30000   .08333   .0335
```

[H] ENTROPY IN CHARACTER 1 = 1.82257

[I] CONDITIONAL PROBABILITY DISTRIBUTIONS   [J] CONDITIONAL ENTROPIES
```
          I(1)      I(2)      I(3)      I(4)      I(5)
J(1)   1.00000   0.00000   0.00000   0.00000   0.00000                -0.00000
J(2)   0.00000   1.00000   0.00000   0.00000   0.00000                -0.00000
J(3)    .02222    .57778    .37778    .02222   0.00000                 1.23189
J(4)   0.00000   0.00000    .14286    .57143    .28571                 1.37876
```

[K] INFORMATION COMMON TO BOTH CHARACTERS = .73779

[L] ENTROPY REMAINING IN CHARACTER 1 AFTER OBSERVING CHARACTER 2 = 1.08478

[M] FRACTION OF INFORMATION IN CHARACTER 1 ALSO CONTAINED IN CHARACTER 2 = .40481

```
   PROBABILITIES OF THE STATES OF CHARACTER 2, GIVEN THE GOOD OBJECTS
        J(1)     J(2)     J(3)     J(4)
       .08333   .05000   .75000   .11667
```

ENTROPY IN CHARACTER 2 = 1.18773

CONDITIONAL PROBABILITY DISTRIBUTIONS   CONDITIONAL ENTROPIES
```
        J(1)      J(2)      J(3)      J(4)
I(1)   .03333   0.00000    .10667   0.00000                      .65002
I(2)  0.00000    .10345    .09855   0.00000                      .47983
I(3)  0.00000   0.00000    .94444    .05556                      .30894
I(4)  0.00000   0.00000    .20000    .80000                      .72193
I(5)  0.00000   0.00000   0.00000   1.00000                     -0.00000
```

INFORMATION COMMON TO BOTH CHARACTERS = .73773

ENTROPY REMAINING IN CHARACTER 2 AFTER OBSERVING CHARACTER 1 = .44004

FRACTION OF INFORMATION IN CHARACTER 2 ALSO CONTAINED IN CHARACTER 1 = .62117

FIGURE 43. Printout of the comparison of characters I = 1 and J = 2 made
            by CHARANAL.

Entropy. The concept of entropy was first developed as a quantitative formulation of the second law of thermodynamics. Entropy is a measure of the disorganization of a closed system—so, the higher the entropy is, the lower is the amount of work that can be done by this sytem. Accordingly, the second law of thermodynamics becomes, in the formulation of Boltzmann (1896), a closed system can vary only if its probability (entropy) is increased by the variation.

The transfer of the concept to information theory was gradual. The first step was accomplished by Boltzmann who found the entropy to be proportional to the logarithm of the number of alternatives possible for a closed system, when all the known information has been recorded. In other words, the entropy is proportional to the logarithm of the amount of information that is missing.

This concept was developed by various authors, of which Shannon (1948) derived from the premises of the theory of information, the equation:

$$H = -\sum_{i=1}^{n} p_i \ \log p_i$$

where H is a measure of the uncertainty, or choice, and p_i is the probability of the various events i or the frequency attached to each piece of information i. He noticed that his equation was similar to the equation of entropy of Boltzmann (1898, pp. 219-221) and concluded that H corresponds to the entropy of an information system.

The output of the CHARANAL program is:

1. Upon request, the form of the initial data will be printed out for recording or editing purposes.

2. The main output in which either all of the descriptors in the study will be compared, one with the other, or any chosen subset of descriptors will be analyzed. The main printout contains:

A. The ordinal number of the descriptors being analyzed.

B. The number of states in each descriptor.

C. A measure of the difference between the two descriptors.

D. A measure of the similarity of the descriptors.

E. The number of objects not used in the calculations because of missing data.

F. The probability number of E.

G. The probability distribution of the objects over the states in descriptor (character) I = 1 to n.

H. Entropy in character I.

I. A matrix of the conditional probabilities given that the objects are in the respective states of descriptor or character and with respect to character I.

J. The conditional entropy for the above comparisons.

K. The information held in common by the two descriptors being analyzed.

L. The confusion (entropy) in I given character J.

M. The fraction of information in character I contained in J.

The same information is repeated in the output immediately following for the reverse comparison; namely, for the comparison of character or descriptor J with descriptor I. Obviously, one descriptor may be dependent on another or both may show a high degree of correlation, but in turn, both, while independent, may be dependent on a third descriptor.

3. At the users option, two additional outputs may be requested. Both show the total amount of correlation of a descriptor with all other descriptors in the study. In one total the actual amount of information held by the descriptors is used and in the other, the amount of information is normalized.

Morisita-Ono. The Morisita-Ono Index of Faunal Affinity is an index for measuring faunal similarity between two communities or between two samples. For every two samples compared, there is one index value between zero and 1.10. If there are many samples to be compared, values are obtained relating each sample to every other sample. Values close to one indicate that the two samples are from the same community; values close to zero indicate that the two samples are from different communities. Intermediate values must be interpreted by the observer. For instance, if several intermediate values are obtained from a sampling grid, an investigator may attempt to fix the boundaries separating the communities.

Computation of the Index. This index was originally developed by Morisita (1959), and rewritten by ONO (1961). It uses the formula for computing the index as shown in Table 2.

The output (Table 3) consists of the code number of each pair of the localities compared and the respective index number. It is a useful index though one cannot assign confidence intervals and the results are questionable if some of the samples are small.

Biological species diversity. This module calculates the entropy or confusion in the data available about any given collection of elements from which a random sample can be taken containing a known number of species or classes of these elements. The output (Figure 44) consists of printing the sample number; the number of species for the respective sample; the maximum entropy, H (MAX); and the actual entropy, H PRIME.

This program determines the amount of unavailable information or species diversity, H PRIME, utilizing the Shannon and Weaver equation (modified by Basharin) below.

$$H^1 = \frac{C}{M} \left(M \log_{10} M - \sum_{k=1}^{i} N_k \log_{10} N_k \right) - \frac{(S-1)}{(2M)} ,$$

given $S = i$ species containing N_k = number of individual specimens (elements) where

$$M = \sum_{k=1}^{i} N_k ,$$

and C is a scale factor for conversion of log base 10 to another base.

Table 2. MORISITA-ONO INDEX OF FAUNAL AFFINITY

$$C\lambda = 2 \sum_{I=1}^{\infty} \frac{(N1I) \quad (N2I)}{(\lambda1+\lambda2)(N1)(N2)}$$

Where:

$C\lambda$ = The index value

$N1I$ = Number of individuals of species I found in Sample 1.

$N2I$ = Number of individuals of species I found in Sample 2.

$N1$ = Total number of individuals in Sample 1.

$N2$ = Total number of individuals in Sample 2.

$$\lambda1 = \sum_{I=1}^{\infty} \frac{N1I\,(N1I-1)}{N1\,(N1-1)} \quad \text{For Sample 1.}$$

$$\lambda2 = \sum_{I=2}^{\infty} \frac{N2I\,(N2I-2)}{N2\,(N2-2)} \quad \text{For Sample 2.}$$

Table 3. MORISITA-ONO INDEX

Affinity Index of Trawl Samples Collected at DWD Site-A and Site-E

12&	13 =	.824
12&	14 =	.847
12&	15 =	.889
12&	16 =	.880
12&	18 =	.001
12&	30 =	.133
12&	33 =	.045
12&	49 =	.081
12&	54 =	.123
12&	7116 =	.017
13&	14 =	.861
13&	15 =	.854
13&	16 =	.275
13&	18 =	.000
13&	30 =	.030
13&	33 =	.026
13&	49 =	.133
13&	54 =	.040
13&	7116 =	.043

Currently, two other generalized modules are being worked on, and both depend on ENVIR for the partitioning of the data bank and the forwarding of the requisite information in the partition to the modules. The first module will provide commands permitting individual and varied environmental models to be performed without having to program for each separate model. The second generalized module will contain the necessary commands to execute the various statistical procedures required for an analysis. This last module will obsolete the 20 odd statistical routines currently appended to ENVIR as separate modules.

In conclusion, ENVIR is a computerized system for the analysis of data. EDMPAS includes modules for the further manipulation, reorganization and display of selected ENVIR-derived subsets of information. EDMPAS is not a generalized data base management system. The latter are characterized by being optimized for a specific requirement and for conditions in which only a few key descriptors need to be addressable. EDMPAS is independent of the nature of the data base, has all fields addressable, but requires that the user be concerned by the scientific content and implications, explicit or otherwise, of the data bank. In this respect, it is unique.

DISCUSSION

COBB—I would like to start off the discussion by going back to a question raised this morning by Dr. Helms when he talked about the idea of having a subset or a mini data base and use of that in the anlysis problem. I would be interested in your feelings, Dr. Fleming, on the idea of using a total data base rather than using a subset of it to determine trends.

FLEMING—Yes, we do that to find trends, as well as for a number of other reasons. Before even one word of ENVIR coding was put on paper, it was determined that any scientific retrieval system had to be able sine qua non to partition the data bank into any desired subset ranging from the null set to the universal set. It follows then, that in an ENVIR command such as SORT AND PRINT, the subset of interest must be specified and this subset may be of any specified dimension or size.

Secondly, if we wish to ask a large number of questions, or do many manipulations of data involving only one part of a data bank, such as some data pertinent to the subset, ENGLAND as contrasted with a data bank of EUROPE, the suitable ENVIR commands will segregate the selected subset and provide the researcher with an individual DYNAMIC FILE. He may use this mini file in isolation from main file as long as he wishes.

Incidentally, I should mention that if the researcher does not consider an independent mini file desirable, but, nonetheless, wishes to ask a series of questions only involving some particular subset such as ENGLAND, he may use the ENVIR command HOLD=TRUE and all subsequent queries will only refer to that subset. The command, HOLD=FALSE, will reopen the whole bank.

Thirdly, a module such as TG4DA can construct an extrapolated, itensified DYNAMIC FILE from a small subset extracted from the master DYNAMIC FILE. In other words, a dense, dynamic file of physical oceanographic data around the Hawaiian Islands from a data bank of data of the Pacific Ocean.

Fourthly, the SELECT DESCRIPTOR command can be used to build any desired DYNAMIC FILES from raw data. We have used this command to construct separate DYNAMIC FILES of biological and chemical data from an initial file containing a mix of both.

SAMPLE NO.	SPECIES	TOTAL IND.	H(MAX)	HPRIME
12345	6	122	2.584962	2.028915
12346	3	100	1.584962	1.490000
3P120	4	100	2.000000	1.985000
89A36	4	100	2.000000	1.985000
3P120	4	100	2.000000	1.985000
89A36	4	100	2.000000	1.985000
ZC121	4	100	2.000000	1.985000
XYZ99	4	100	2.000000	1.985000
3P120	4	100	2.000000	1.985000
89A36	4	100	2.000000	1.985000
ZC121	4	100	2.000000	1.985000
XYZ99	4	100	2.000000	1.985000
3P120	4	100	2.000000	1.985000
89A36	4	100	2.000000	1.985000
ZC121	4	100	2.000000	1.985000
XYZ99	4	100	2.000000	1.985000
3P120	4	100	2.000000	1.985000
89A36	4	100	2.000000	1.985000
12345	6	122	2.584962	2.028915
12346	3	100	1.584962	1.490000
ZC121	4	100	2.000000	1.985000
XYZ99	4	100	2.000000	1.985000
3P120	4	100	2.000000	1.985000
89A36	2	50	1.000000	.990000
+36	2	30	1.000000	.633356
ZC121	4	100	2.000000	1.985000
X0Z99	1	25	.000000	.000000
XYZ99	3	75	1.584962	1.571629
3P120	4	100	2.000000	1.985000
89A36	4	100	2.000000	1.985000
ZC121	4	100	2.000000	1.985000
XYZ99	4	100	2.000000	1.985000
3P120	4	100	2.000000	1.985000
89A36	4	100	2.000000	1.985000
931 1	4	80	2.000000	1.804442
XYZ99	4	100	2.000000	1.985000

FIGURE 44. Biological species diversity printout.

Lastly, and this may not respond to your question, but it is a type of subset, ENVIR can be asked to print out the first 5 or 10, etc., lines of information in a response. This facility may enable the researcher to determine whether the full output is worth having.

PEACHEY—Mr. Chairman, when this meeting was first discussed, what interested me was the way in which Dr. Oppenheimer and Dr. Fleming emphasized the integrity of a single, individual measurement. And I think that is really why I decided to come here. One could see that in the space program an individual measurement was crucial even if it was buried in a mean value for normal measurement purposes, it could be recovered when something went wrong. It seems to me that it's this capacity to oscillate between detail and summary in a mathematical or intellectual sense which is the prime requisite we need to concentrate on.

We were talking this morning and wondering whether perhaps if we met again in another five years or so that we might pretend that we were part of somebody elses management and say, "Look you, stay in your office and we hope it's a big office and you have all of the scenes from these individual measurements to what the boss at the top wants to know and is always saying, 'don't tell me the facts, show me the picture'."

But, I think we shall never get straight until we articulate the whole series of processes from data capture to decision and stress the importance of individual records and of resynthesis and reanalysis of these records as the decision-making process goes on.

OPPENHEIMER—I agree thoroughly, Dr. Peachey. The environmental analyses of the future will depend not only on a pre-synthesis of data but also on a reconstruction or updating of individual environmental data points that may provide new synthesis or meaning.

BROGDEN—I was thinking that one might be able to conduct an experiment to see if subsets would really be adequate by putting someone in a simulated decision making situation and see how well they do with a subset of the actual data. Perhaps you could graph success in making the correct decision against the size of the subset and find out if a certain size of subset is entirely adequate for certain types of decisions.

EVERHART—I hesitate to get into this because we just started this process about three months ago, but within the Chesapeake Research Consortium we have in the EDGES Program, a group called the case study group, which is working in concert with the decision making agencies on Chesapeake Bay in viewing case studies. There are eight of us including Lyle Crane from Michigan, Bob Byrne from Virginia, and Bob Ellis from Connecticut. The rest of us are from the upper Bay area.

I am sorry I didn't bring a copy of our first case study with me. It is a study of a small creek in Annapolis which the Corps of Engineers has found to be a very sensitive area. What we eventually came out with was a case study which has been published. The Corps of Engineers has attached a letter to it, has sent it to the State Department of Natural Resources and the City Planners of Annapolis, and has urged as a result of this case study (which fell out of the Research and Management Shoreline Data Bank) that a moritorium be imposed on the waterfront development of the entire Back Creek region until a comprehensive plan is developed. This is a case where a decision-making agency accepted a case study, which we offered with some trepidation, without reservation.

What we had done with the computer is that one day we looked at the Back Creek area in Annapolis. There are currently about 1,350 boats berthed on this one-mile-long creek, and we estimated that with about another 500 or 600 boats the creek was going to be congested to the point where people would not want to put their boats there. And so the computer was queried to determine how many applications existed. We specified waterway code and grid coordinates, and lo and behold, we came out with eight active applications on Back Creek. If you keep in mind that the number of mooring piles is roughly equal to the number of boat slips a person is asking for, and if you add them all up, I believe there are about 400 slips that were applied for. On the basis of this, we did the case study for the Corps of Engineers and the Maryland Department of Natural Resources.

We are currently into our second case study concerning the Baltimore Harbor dredging. And the third case study that has been requested of us is of the development in Dorchester County, Maryland. The sort of study that we have been doing has completely reshaped our computer information system. It is just a never ending circle, you go around and around. But I thought I would bring this case study out because it speaks to the very thing I think you were discussing.

PEACHEY—Mr. Chairman, may I give an example of something very important. If your bank manager sends you a summary of your mean expenditure for the month, it would be interesting but you would have lost the facility of seeing where you present position is because you can't tell which checks have been returned for payment. And I think we have to consciously resist the temptation to consider the premature aggregation of data; whereas, we have all been trying to do the opposite. We are trained to eliminate detail in order to produce a readable scientific paper and to move to subsets and means as quickly as possible. Now I think we ought to be experimental in this area and show as much concern for data as we do for payment slips. I just can't see the difference in values.

Constantly in real life operation situations we do have a set of non-aggregated data and we accept it and we produce from it some kind of perhaps verbal or written or non-arithmetic summary, and I sometimes think certain biologists and their obsession with the whole concept of mean value and normal distribution is perhaps a big mistake. Today's noise may be tomorrow's signal, as Dr. Oppenheimer pointed out.

ROSENFELD—Dr. Fleming, your data seem to all be classified; don't you use any raw data? The way you described it, it seemed to me you have already put the data into a classifications system.

FLEMING—The ENVIR program usually contains the rawest kind of data, but you use it to select states of taxonomic significance or interest for input to GRAPH. Most of this sort of data that would be input to GRAPH does not lend itself to statistical analysis, because most of it is qualitative and very little of it can be ranked in any fashion or assigned any sort of continuous numbers. If we had data suitable for statistical analysis, we would direct ENVIR to input to the selected or appropriate statistical sub-routine.

ROSENFELD—Do you do trend surface fitting in your mapping?

FLEMING—Yes.

ROSENFELD—You do use such fitting for your interpretations?

FLEMING—Yes.

BROGDEN—In talking about these techniques for mapping from badly distributed sample points, I was wondering what people's experience with contour mapping was. We tried a technique one time to fit polynomial surfaces to badly distributed points and use them to produce a contour map, but toward the edges of the field of data the contours got very poor and I was wondering if any other people have experience in this contouring effort. What sort of contouring approaches work well? Mike (Ellis), I understand you use a system called SYMAP. What is the basis of that, extrapolation?

ELLIS—It's purely linear, a function of distance.

BROGDEN—So you just use a simple interpolation?

RAUSCHUBER—The Board has several different contouring techniques available for use on our UNIVAC 1106 computer system. In addition to the SYMAP routine which Mr. Ellis previously described, we have a contour plotting package called CPS-1. This contour plotting system provides a software package which is designed to plot contours, draw cross sections, spot elevation points, compute areas and volumes for cut-and-fill, and perform many related operations. The TWODB also utilizes a three-dimensional plotting routine, which allows the user to plot an object as though being from any angle the user chooses. One problem we have experienced with the SYMAP and CPS-1 package is that of spacial distribution. Frequently, the package will attempt to plot outside the defined boundary of the data subset.

ELLIS—In the SYMAP system they allow you to impose boundaries which impede the extrapolation process, such as a spoil bank.

BROGDEN—To prevent it from extrapolation across that boundary?

ELLIS—You give it a percentage, or the weight of the effect of the boundary.

LOUDON—One of the problems of contouring, of course, is the problem of traverse data. The data are closely spaced along tracts with large gaps in between.

This is handled reasonably well I think in various commercially available packages such as the SACM package, developed by ACI here in Houston. Values on a rectangular grid are interpolated from the data points. Date are considered in each of eight octants around the grid points, and equal weight given to each octant, regardless of the number of data points in each. That overcomes the tracked data problem to some extent but I don't think any of these packages satisfactorily overcome the problem of what to do with large, empty gaps. This is something which has to be decided by the scientist.

One way of deferring a decision is to use a contouring package which leaves areas blank where there is poor control. Another technique is to use iterative-fit, trend-surface analysis. A global polynomial if fitted to the entire surface and where there is good local control, local polynomials are superimposed on the global fit. The resulting map is of a surface that is highly variable in areas of good control and smooth elsewhere.

ROSENFELD—That is what you do with it, Mr. Fleming, in the EDMPAS system?

FLEMING—Yes.

ROSENFELD—You analyze the anomalies?

FLEMING—Yes.

RAUSCHUBER—I would like to ask Dr. Fleming a question. As clustering analysis is a relatively new "tool" in that it was developed within the past 25 year period, have you tried similarity testing using different type clustering techniques?

FLEMING—Yes we have, sometime ago. And the technique we have settled on seems to give us the best correlation with biological reality. Because what happens in our clustering program is that a new cluster is formed as you lower the similarity and you know which object has caused the cluster to change and from a biological point of view you can ask why did that object go into the cluster at this particular similarity value and you look for a reason.

RAUSCHUBER—We use something very similar in our estuarine studies for evaluating the response of white shrimp productivity with various environmental parameters. We have had good luck with hierarchial type clustering which uses the distance between group means.

FLEMING—I will give you one example, I think an actinomycete was involved. I am not quite sure now, but anyhow a chap from the research section of a drug company came into the office with data on strains of these antibiotic organisms. It was typical bacteriological sort of data, dealing with type reaction, but with some morphological data. As he looked at the clustering sequence in the printout, he turned white because the clustering sequence duplicated the sequence of genetic crosses that they had performed and this was, of course, proprietary information. The clustering followed the geneological, if not phylogenetic, history. We had no knowledge of this procedure.

RAUSCHUBER—I would like to ask Dr. Fleming one more question. I live in two different communities, one that backs clustering techniques, and the other that backs factorial type techniques, factorial analysis. Dr. Fleming, have you dealt with factorial analysis?

FLEMING—Yes, one of our former members of the staff worked with factorial analysis. The difficulty there is that most of our biological data is qualitative and it is difficult to assign numbers that are meaningful in any way. As a matter of fact, I have several of those programs lying in my closed files, and I haven't used them in a long time.

COBB—On this same subject our own work looking at data base systems has led us to feel that the type system that you use is very dependent on the type of data that you have. I am wondering that with the number of different situations where you have used the ENVIR system, how you have found the effectiveness of the ENVIR system to be. Is it very dependent on the data that you have?

FLEMING—ENVIR does not care what kind of data you use, We have had data banks of legal, geological, socio-economic, physical, chemical, biological, soils, pesticides, medical, and so forth. In response to your questions, however, commercial programs designed for business data operations involving income tax deductions, social security deductions, accounts receivable, and so forth, would handle that type of data better where the operations are predetermined and repetitive, and a generalized program is not required and possibly not even desirable.

However, since that type of program· is specialized and designed for specific operations, they are unsuitable in areas requiring investigation and research. In the latter, the operations cannot be predetermined, for no other reason than that the end product is unknown. ENVIR has been designed for science and logic. Logic not only in the formal Kantian sense, but to enable the investigator to discover the meaning of his data, and then the implications of the meaning.

Let me close with some figures. If we use only one command of ENVIR, the SORT AND PRINT command, we are able to specify the number of subsets of a bank equal to the result of raising 2 to the power of the number of records in the bank. In other words, only 20 records enables us to designate more than a million subsets. This same command allows us to specify what information to print out for each subset, arrange the information in any order, hierarchically arrange it and format it. The result is that the number of possibly different printouts given by only 10 records and 3 fields or descriptors is over twenty-four and one half billion, a ridiculously enormous figure, but mathematically correct. Obviously, one will never ask so many questions, but a scientist must be given the freedom to ask what he wishes.

EBERHART—It seems to me that almost any system you operate under, whether it's System 2000 or ENVIR or GIS, can provide you with this sort of information if properly structured. I wonder if you would comment on what qualities you believe ENVIR has that no other system has.

FLEMING—One thing is the compression which is at least one order of magnitude. The second thing is that we have not inverted the files, but we have the same results as having all the files inverted. ENVIR arithmetically calculates the location of the desired information. The other systems must record the addresses of the inverted files with the result that as you invert files, the addresses grow geometrically, and soon exceed the memory requirements of your original data. You can get the same results with other systems as with ENVIR if you have a large team of programmers on hand to keep changing your files for you and you are willing to wait for them to do it. ENVIR just does it faster, more efficiently, and with the scientist in control, rather than a computer personnel team.

OPPENHEIMER—The term "compressibility" is one of the subjects that will be very important to discuss at the end of this meeting. If we extrapolate our on-line data capabilities into the future and the past, we must have a compressed system. It's one of the absolute necessities. Otherwise we will find ourselves with so many data points that we will be back to our present position, that the data are in the literature and when needed must be searched out.

KOHNKE—Dr. Fleming, I have a question concerning the documentation related to your analytical values. Besides the quantitative values of the parameters, it is very important to know how they were recorded or analyzed. This additional information is necessary for the explanation of the data and the comparability among data banks. Could you tell us something about this documentation in the ENVIR system?

FLEMING—Yes, unfortunately, this wasn't a Conference in which I was to talk about ENVIR and that takes another hour or so. For the units in which any of your measurements are made, there is a provision for a label to be fastened on to that field or descriptor, and whenever you get a printout, it will print out whether the measurement was parts per thousand or parts per million, and so forth. If you want to know what sort of instrument was used, there is a memo command that can be used that will print out just what instrument was employed in doing this and that.

OPPENHEIMER—It will print out the source of the information in the bibliographic file so that one can go back to the source.

ROSENFELD—I confess I got lost somewhere in the beginning. How is this compression allowed?

FLEMING—That's another hour's discussion.

ROSENFELD—Well, I mean just briefly, what do you compress the raw data into?

FLEMING—The compressed dynamic file is a two-dimensional matrix. Each column in the matrix holds one complete record. The first field in all the records will be placed in the uppermost bits of the column. The remaining fields will follow in sequence downward. Let us assume that the first field is an alpha field. We do not need to consider whether we have 3 records or 3 million records. We are concerned with the number of states that will occur in the first field for all the records. The unique states may be one character long or 130 characters long. Each will be represented in computer core by a unique binary configuration or number, automatically determined by the program as the data is read. The number of bits needed to represent all the states in the field is equal to the power that 2 must be raised to contain all the unique states. Given 3 or 3 million records that for this first-named field only assume the states, red, white, or blue, then the two first bits in each column will store this data. All zeroes in the binary representation will always mean "no information" for the respective field. State red is represented by 01, white by 10, and blue by 11. A single dictionary is automatically built containing the code and associated alphabetic data. It follows that 8 unique states require 3 bits; 16 states, 4 bits; and a million states, 20 bits. In a coded descriptor or field, the user assigns a decimal number and its alphabetic equivalent before building the bank. Numeric information is handled in the same way, except no dictionaries are necessary. The latitude and longitude of the world to approximately six-foot accuracy is handled in 38 bits per record. The program calculates arithmetically the location of the bits representing the requested information and sorts using these compressed fields. The dictionary of the name and coded fields are only used just prior to printout.

ROSENFELD—So it's a large dictionary?

FLEMING—It could be, but rarely is for scientific data. Commercial operations might have client's names and addresses. Each would be unique to a record, and, consequently, the dictionaries would be large. One does not do logic on such data. In scientific data, things can only assume a limited number of shapes, have a limited number of colors, etc., aside from the bigger consideration that many fields in scientific data have some implied correlation possible which means that the states cannot be unique over the records. Much scientific data is numerical and this data does not have ENVIR dictionaries. Over a trillion unique numerical states for over a trillion records would not have a dictionary, but would be contained in only 30 bits of each record.

By the way, before we wrote one line of coding, we anticipated scientific data of the name and address type, such as precise locality, as five miles from Smith's Ranch by the granite boulder under the oak tree. We wrote the algorithm for going out to random access with such data, but we have never had occasion as yet to implement it. We will eventually.

BROGDEN—I would like to show you something that ENVIR does not do. While I am working with ENVIR all the time there is a problem, and it has to do with a type of data representation that we get into. Let's say that I had a data file of stations in an estuary or something like that for which various parameters have been measured in it. I am going to represent this as a tree structure. In other words under a station name I have parameters. In this case, A, B, C, and D, and in the other case I will have A, C, D, E. These are the names of the parameters that have been measured. In a system which represents a tree structure, like System 2000, you can ask a question or you can request all the stations which have both parameter A and D measurements, but in ENVIR, ENVIR will represent this as straight line, rather than a tree.

I have been experimenting with representing this kind of data and I ended up using just a single parameter under each station and the numerical values that came with that parameter. So to represent these four parameter measurements I would need four items essentially and I can't ask certain questions. With a tree structure, you can say let me have the data which has A and D measured, and you can't really say that in ENVIR. It has to do with the structure of your data, a tree structure compared to a strictly linear structure. In other words, in an ENVIR item each descriptor can only have a single state. There can only be one name per parameter, there can only be one genus or one any other descriptor there.

There are ways to get around this and I get around this by asking this question essentially separately, use ENVIR to sort out all of the A's and all the D's and the program sorts everything completely. They come out next to each other in the output stream and then I simply ask another program to match them up. So it is something that ENVIR doesn't do but it's something you can get around.

We were talking about sharing data through some sort of a total system, I don't know how we are going to tackle the problems of representing the data structure. There may be some mathematical shorthand or topological shorthand, some sort of mathematical way to represent a data structure, but this is something that has to be available before you can use data from another source, you have to have an idea of the data structure. And I don't know how we get anybody to define a mathematical or graphical method for representing data structure.

KOVACS—What is the relative machine independence of this particular system?

FLEMING—It's all written in the FORTRAN.

KOVACS—On a UNIVAC?

FLEMING—Originally on a CDC 6400. It is now working on UNIVAC, and limited versions on IBM 360-370 series and Burroughs 6700.

KOVACS—Is the system proprietary or is it being made available?

FLEMING—No, it's proprietary for GURC people. We had a clearly defined objective when the system was developed which was that the scientists would interface with the computer. The scientist designs his own files, creates his own dimensions. Given 3 hours of training any scientist with no computer knowledge whatsoever can create his own data information file. We were also concerned with absolute file content addressability and this was also achieved. These were the necessary ingredients in the system.

CUTBILL—Can I say something to this business of the competitiveness between generalized systems? It is very clear from the talk that is going on today, that there are no general systems, nobody has them. And it seems to me there are four things we've got to look at here. First of all, the system obviously must accept and preserve the information you want to store and most systems that have been talked about around this table, from my point of view, fail on that because I have this type of prestructured information which would be very difficult to put into ENVIR or something like this. You may be able to apply the data, but I would guess it's not easy. GIS does it all right but you pay a lot of dollars for it. So there is that kind of problem.

Then the second problem I see, which many of the systems do cope with rather well, is the business of being independent of the generality of the data organization. Obviously the detail of the data matters a lot. If you are trying to run a statistical regression program, the data you throw at it has to have a certain rather precise form. But how those data appear generally doesn't matter. This is the basic function of a data management system and most of them seem to have this.

The third problem which is often neglected is the good IO. Data must be easy to get in and out of the system and again most of these systems are usually very easy to get in and out of for the data they represent. However, the moment you start pushing them a little bit, all of a sudden the IO gives you trouble and you find it is necessary to write preprocessing programs and this sort of thing, and you've lost a lot of advantage.

The fourth problem is quite simply the range of capabilities. All these clustering and mapping programs and things are just extra things you add on so that you can call up a system that can do the job in hand.

Now the thing which concerns me the most is that I want access to all these different capabilities. When I've got a kind of high speed retrieval problem, I want to be able to swap data into ENVIR where it can be handled and then return the material to my original system where I can do all my text handling and type setting.

So we've got a problem of how we move data from one system to the other. There is work going on in this problem of describing data. I can't quote the exact reference, but I refer anyone who is interested to a series of papers in the IBM Systems Global Journals (about a year or eighteen months ago) by a man called Senko. He has been doing a lot of very detailed thinking about generalized descriptions of the structural properties of information which are totally independent of how the data is represented. It's obviously an area in which we have to do a lot more work because any total system must go as a complete self-describing package with standard descriptors. One must approach such a system by program and then break it down to reformat it into a conventional sort of receiving system.

FLEMING—Our input to modules will put out on tape the information that is necessary to feed into any models that you have and you could fit it into any retrieval program. We have developed a pre-ENVIR module that will process fixed-field data designed for other programs, for reception by ENVIR. We are continuously doing this with EPA and Navy data.

In addition, once we have a number of pertinent dynamic files built, even though they are disparate, we can correlate the data contained in each dynamic file in the same runstream. For instance, we have map contoured Lead concentration by dotted lines from a chemical file, sediment type by a solid line from a geological file, and depth by a dash-dot line from another file, all on the same paper in the single runstream.

BROGDEN—You have been adding things to your program since I last got a copy.

FLEMING—There have been a number of things added. One very useful addition has been the ability to partition a dynamic file and create a new independent dynamic file from the old file. We can select fields from two or more separate dynamic files and create a new dynamic file to be queried. We also are able to process fields from a dynamic file with appropriate modules such as Typographically Guided Four Dimensional Analysis module and insert the new fields created back into the old dynamic file as new, additional fields.

And in reference to the difficulty of data structure that you were referring to before, Bill (Brogden), we use this for restructuring a dynamic file. This has been particularly useful to us in our marine acoustical work where we have created a number of differently structured and oriented dynamic files from the original dynamic file.

CUTBILL—You are describing a computer trick doing all the science.

FLEMING—Right.

OPPENHEIMER—It seems to me we want everything, don't we? We want to be able to put our real data in to keep it, use it, extract it, and compress it. We want to be able to have all of the terminal peripherals—to have the graphing, the clustering, the mapping, the contouring, etc.

The question is, today we don't have a general system, what do we do? Will this be one of our recommendations? If we are going to have a data commonality, then we have to describe what we want in the system. And so we are now beginning to see in this session some of the computer aspects that we would desire in the final recommended system. The system described by Mr. Fleming seems to be one general system that might serve as a starting system.

KONKE—I would like to make some remarks on a general data system which has been developed for exchange purposes in the GARP Atlantic Tropical Experiment (GATE). As Dr. Austin already mentioned, this is a multidisciplinary experiment which will take place from June 1974 until October 1974. The whole experiment only makes sense if the meteorological and oceanographical data can be exchanged among the participants and be jointly analyzed. To assure this, a general data exchange format had to be developed. Due to the varying computer facilities available in the participating countries and due to the different observation methods and recording techniques, the exchange format had to be as flexible as possible. A great deal of data exchanged in this format has a documentary character, i.e., it is additional information describing the recording technique, the quality and the units of the parameters, the analytical method, the tape format, the code character by character, and a lot of other things.

It's true, a general data format applicable to a great variety of scientific disciplines requires a very complex system. Therefore, such a system should only be used for an exchange of data; it is absolutely unsuitable for data processing, because it is extremely costly and time-consuming. Today, I cannot report about the success or failure of the GATE exchange format. But I would think that it is a first approach to an international general exchange format, which, in my view, could form the platform for a worldwide intensified communication among data centers or institutes.

ELLIS—My understanding of your system, Mr. Fleming, and I have talked about it at some length with Dr. Brogden and some of your staff, is that the compressibility is very dependent on your ability to estimate the number of descriptor states before you begin to load the data. Is this the case?

FLEMING—Yes, it's not a critical thing because at the present time with UNIVAC we can re-dimension any inadequately dimensioned fields at any time during the life of the dynamic file.

ELLIS—Well, our experience at the Water Development Board, especially working with Mr. Nelson and Mr. Rauschuber, is that for an on-going system which is continually having data added to it, such as our Coastal Data System, things change too much over time and we cannot possibly guess or estimate, in the early stages, exactly how many different states we will have. Therefore, it concerns me that you probably have to recreate your data base if a descriptor gains states. The way I understand your structure is that you allocate just enough bits for a descriptor and no more.

FLEMING—The answer is no. We do not have to recreate the original data. In order to use computer memory with maximum efficiency, we only allocate the number of bits required by a particular data bank. However, the command DEFINE MORE DESCRIPTORS automatically increases the size of the compressed file.

Incidentally, if it is necessary to increase the size of any field within an existent compressed file to accommodate some unanticipated parameters in some new, additional records, the command REDEFINE DESCRIPTORS is used. In this command, one identifies the field by name and states the desired size. All the old compressed records of course remain as they were and one just proceeds to add new records.

ELLIS—I would also like to make a couple of comments concerning generalized versus specific systems. At the Water Development Board we have many different types of data. We have very seriously considered both types of systems and at the very beginning I was in favor of designing a system that would handle any kind of data. However, we opted for an approach which used a large number of different systems, and when I say systems I mean computer programs, each of which handles a specific kind of data. It is our feeling that a generalized system will handle any kind of information fairly well, but no information exactly right. So what we have done is take an approach of designing a monitoring system which watches over all our other computer programs. It's a CAI-type program that comes in and says OK, we have our—let me back up a little bit.

We have decided to categorize all of our data into five categories of natural resources information in the State of Texas. The Monitoring program comes on and asks which of these major categories of information are you interested in. (I was going to demonstrate this to you but we didn't have time.) After it goes through that and you finally pick a file number, it starts querying you as to what options you would like to use; i. e., how you want to see the data, etc. All it really does is prepare another run stream. If you are familiar with the UNIVAC you know that you can dynamically initiate a batch program. The Monitoring program does this and then allows you to begin another request or check the status of a request, etc. So we can pull together all our different systems under one house, and that's why we can account very precisely for who has retrieved what data. From this information we can determine what data to keep on-line to the computer, etc. Therefore, at the Water Development Board we have taken a slightly hybrid approach to specific versus generalized systems.

LOUDON—With some geological data we are faced with a number of data files that have rather different properties. And in designing a system to handle any of these one is faced obviously with a trade off between such things as compactness of data storage, rapidity of access, ease of handling the data, output from the data management system, processing programs, things of this nature.

I would like to outline one approach that we are taking at the Institute of Geological Sciences to provide a generalized system for data handling (Jeffery, 1974). All of the processing programs we are dealing with handle information in a two-dimensional array. Maybe this is a historical accident, but I think there are reasons for supposing that a matrix is a particularly suitable structure for handling data. Therefore, most of the data sets come out of the data management programs, no matter how they are stored, as two-dimensional arrays to be passed on to any other program within the system. This is the internal exchange format. Of course, any other structure can be mapped into a two-dimensional array although it may result in a very sparse matrix. If one is coming out from one system on one machine into another system and another machine then the two-dimensional array seems a very convenient way in which to pass those data across.

Of course, there are many methods for compressing data for purposes of transmission or whatever, and then expanding again before processing. Where we have a number of different data files dealing with different sets of data, greater generality can be achieved by means of the header records as described by Dr. Kohnke. They describe such things are the names of variables, how they are formated, the length of each field, how many items there are, how missing measurements are recorded, what dictionaries are required to look up some of the elements of a particular field. The header records can also store the maximum and minimum value within each field to pass on to programs for statistical analysis, for data checking, graphic presentation or retrieval. On retrieval of a subset, the header records are also updated. This, I suppose, is not unlike the data description tables in IBM's GIS system.

The other thing in aiming to achieve a reasonable degree of generality is to put all the programs under one program generator. So from the users point of view, he is controlling the system with a set of simple, predetermined, English commands. He specifies the names of the variables which he wishes these commands to apply to, such as printing, analysis, or retrieval. The program generator does two things, first of all, it calls up the job control language which enables access to these files to be made, and it also calls up the FORTRAN programs which actually process the data. And, therefore, from the users point of view he is isolated; he is one level removed, from the systems programs. This interface is provided, but the programmer is able to write FORTRAN programs and bring them into the general system without having to bother too much about the system constraints. He can write a fairly general FORTRAN program and plug it into the system, where the user no longer has to bother about job control language or FORTRAN.

The G-Exec System, which has been developed in the Institute, has these features and seems to work well up to the point beyond which simple English statements are inadequate; for example, to describe a formula for generating a new value, or the distance between two items as opposed to their actual coordinates, for use in retrieval or for analysis. It is very difficult to generalize at that level, it is difficult to use English and much easier to use a program language with facilities for looping, branching, and a suitable algebraic notation. Therefore, for that purpose, one writes in FORTRAN and incorporates that subroutine, temporarily at least, within the structure of the program generator.

That is one approach that we have taken in IGS to generalization and I think it may enable us also to retrieve data from the system in a way in which it can quite conveniently be passed on to another system for other types of analysis. For instance, it seems pointless to write our own programs for automated contouring. It is much easier to write a program which can extract subsets of data in the form which these other program systems require.

Personally, I doubt very much whether it is possible to get one integrated general system to do everything which is required in environmental science. I think it much more likely that we should be concentrating on the problems of interfacing between systems. So that you can get out of one system and into another.

COBB—I would tend to agree completely with you on this subject. We are able to define the types of attributes that we expect from a data management system. This fact has been under discussion for the last hour or so and our direction should now be towards the development of standards for exchanging data.

I would add another attribute which I think you mentioned briefly; i.e., lack of system support to a data based management system. Some systems require a great deal of very detailed technical support to get the system operating and keep it maintained, others are very simple and very easy for users to use. This fact is crucial to some users and trivial to others, depending upon the size and expertise of their in-house resources.

PEACHEY—I am just trying to learn so that if I ever spend a reasonable amount of time back home, which isn't likely, we could begin to fulfill our mandate to try to pool together national monitoring. I really came here to learn how one might set about this very difficult situation.

I agree with John Cutbill on the possibilities of a non-bibliographic exchange format. And whatever people do in their own ways, you don't need to interfere with the way they tie the stuff in, either manually or in a computerized form. You first get it labeled in a common fashion and you find that you can use it for a variety of other applications as well as exchange. It then seems to me that we really do have to borrow the experience of the commercial world and experiment with unambiguous and unique records which have no structure in them whatsoever, just single measurements with an annotation. Annotations are important, especially where keyboarding facilities are linked to automatic environmental sensors so that "field abbreviations" are attached to data at capture point. You then cluster to create an overview for general management and you retain the individuality of records for specific applications.

One real problem with individual data records is arranging their aggregation and re-analysis across sector boundaries. Again, I think we need to look very closely at the way in which commercial interests set about these tasks, especially where the original data are used for other management purposes not foreseen at the time of original data collection.

It would also be useful to see the extent to which these approaches would replace the present scientific and administrative reliance on expert groups which are almost temporary information systems.

COBB—I think it is a very interesting point you have raised, but I think perhaps some of our answers are already available in the environmental area. For example, we talk about weather prediction. It is clear that the existence of national borders cannot be allowed to effect the weather forecasts of the involved countries. The forecasts must be compatible, if not identical, across the borders. And in order to achieve this objective, it is necessary to exchange data.

This example of data exchange also brings up the issue of summarized data again since a very basic aspect of an individual country's weather forecasting includes the production of both regional and national weather patterns. Such forecasts rely upon different aggregates of the same basic data base.

In short, I believe that we have working examples in our own area of environment data exchanges and we would be better advised to base our development plans on those existing areas. Other examples in the environmental area exist and possibly someone could address the specific issue of data exchange in the marine environment.

KOHNKE—There were and there still are a lot of international oceanographic programs for which scientific data has been exchanged (e.g., CSK—Cooperative Study of the Kuroshio; CICAR—Cooperative Investigations for the Caribbean and Adjacent Regions; CINECA—Cooperative Investigation of the Northeastern Central Atlantic; etc.). But they have always attempted to integrate the exchange of data from such big programs in the rules on the international routine exchange.

As I see it, the exchange of physico-chemical data which are of a quantitative type is generally no longer a big problem. If there are digital data not yet being exchanged, I believe this is mainly due to individual hardware and personnel deficiencies. What we could not solve satisfactorily to date is the establishment of data systems suitable for the exchange and retrieval of scientific results which are of the descriptive type. This is mainly the case in marine geology and hydrobiology. In order to assure that secondary users will come to know about the existence of observations, discipline-oriented inventories have been introduced by which the data centers can refer the user to the observer.

Some years ago an international agreement was achieved for the "International Marine Geological/Geophysical Cruise Inventory" (IGGCI). Since that time, data has been exchanged and the system works quite satisfactorily. Another group of experts, chaired by Dr. Colebrook, came up with the inventory "Results of Marine Biological Investigations" (ROMBI). It seems to me that it is currently hopeless to aspire to a more advanced exchange of data in these fields.

COLEBROOK—I would agree. Some forms of marine biological data are being considered now for exchange and formats will be developed, information like carbon 14 productivity values, which are reasonably standardized. But sample counts for species measurements, this sort of thing, there is a question of exchange of these data as data. Therefore, we have developed an inventory system which would be the only medium of exchange, telling other people what exists and that is as far as it will go for the foreseeable future.

ROSENFELD—I would like to introduce a new variable which none of us has yet talked about. We talk about the geological data and the biological data, chemical data and meteorological data and/or information. We have in geology and biology and in chemistry a level even more basic than raw data and which takes up a great deal more room than the data. I am referring to the actual samples and specimens. In routine weather measurements someone goes out and measures the number of inches of rainfall and throws the sample away, or, in many other areas, there are rapid analyses which destroy the samples.

In geology and biology the question that Dieter (Kohnke) brought up about wrong analysis or interpretation is important. The people as individuals are often very suspicious of other people's analyses or identifications. They are always going back and getting samples and re-doing them and making new identifications; this doesn't occur in many other subjects. We have, if we are trying to compile a data base, different identifications and different data from the same identifiable sample item. This is where we really need identification of not only the method, but also the management of the data base. This is something we haven't talked about at all and this is how you can end up with the same sample in a data base or data bank with more than one identification.

COLEBROOK—I'm sure the biggest international exchange of biological data or information carried on at the moment is the transfer of animals and plants to taxonomists.

ROSENFELD—The actual samples?

COLEBROOK—Yes, the actual samples.

ROSENFELD—The Smithsonian Institution collects samples and the JOIDES Deep Sea Drilling program has all the cores and materials stored. Data extracted from a system have really got to have a great deal of the sample history along with them.

KOHNKE—That's true. I would like to expand on what has been said. The marine geological format we proposed to the Intergovernmental Oceanographic Commission (IOC) exactly meets what you mentioned. The most serious impediment to the wider and more economical utilization of sediment cores and dredged rocks is the lack of an effective means of informing researchers of the nature of the samples in a collection. To fill this need, it was attempted to determine which items of information on such samples are most useful to a wide spectrum of researchers. The marine geological format recommended by the IOC is not so much meant to serve for the recording and transmission of large volumes of quantitative data suitable for synthesizing for special research purposes, but rather is designed to inform marine geologists of the nature of the materials available in the collections.

ROSENFELD—The Deep Sea Drilling program does definitely have hierarchical levels of data and, what's more, the information in the first go-around which gets into a system is apt to be changed with later analysis. For raw data in other subjects that does not happen too much.

OPPENHEIMER—One of the advantages of an inverted file such as ENVIR is that you can add or subtract names; you can even correct names if the species lumpers and the splitters want to change a given genus. It's very easy to correct because you can pull a bit of information out, correct it, add to it, and put it back in. For example, if a new name is created or if preliminary data needs correcting, the file is entered, the data completed, and the file returned. I think that this matter of compressibility and flexibility is the answer to the whole question of environmental data management that we are talking about.

I'm not saying that we change original data, I'm just saying we're editing it.

CUTBILL—We are not changing the name, we are simply calling it about four different things.

OPPENHEIMER—OK, so you can call it four different things, and ENVIR can retrieve any one of the four, but the data base is not changed, you are just adding to and subtracting from the original.

CUTBILL—That's a technical question as to whether one system enables you to do this easily or not. Let's just not go into all that now. It's not relevant to the point that is being made concerning the accumulation of geological or biological data over quite a long period of time and the vital need to have the antecedents of all your data preserved and passed on with it.

ROSENFELD—That would be the literature mostly.

CUTBILL—No, very little of that ever gets published. This is one of the difficulties in geology; people will not publish it for you; they can't afford to.

ROSENFELD—I didn't mean the data analysis.

BROGDEN—This is so you can be sure of associating the later reworkings of the data with the preliminary evaluation. You don't want somebody coming across an early version of the data, not knowing that a later reworking has changed things.

CUTBILL—You never could except by accident.

BROGDEN—I agree, but this would cut down the number of accidents.

ELLIS—I believe that we could define several classes of information and for each one there would be a good way to handle it. Textual information, as far as I am concerned, is one of these classes. There is a program, called GYPSY, that has very good characteristics for searching textual information. It allows you to eliminate prefixes or suffixes, considers the relationship of words in a sentence, etc. It is made to search abstracts and it's very good for that purpose.

Another example that I'll toss out is geographic data. There is, I think, a very specific way to handle and store it.

Then perhaps there are several other classes, but I will lump the rest of them into one class that I call hard data. For this type of data perhaps ENVIR is a good system.

EBERHART—Well, I'm certainly not a proponent of, or knowledgeable in, the system, but there has been a system developed to handle textual information called INFO 360, which runs on either an IBM 360 or 370 series. I don't know much about it, but I can get you information on it.

ELLIS—The problem with GYPSY is it's written in IBM language.

OPPENHEIMER—I would like to ask, Dr. Kohnke, how much your program overlaps with NODC, and can you get on-line with NODC?

KOHNKE—On-line between the two?

OPPENHEIMER—Yes, between the German Oceanographic Data Center and the U. S. one, and perhaps we could include the English one as well.

KOHNKE—Well, no, frankly that's impossible.

OPPENHEIMER—It is?

KOHNKE—The way to receive data is usually asking by letter, and receiving magnetic tape printouts, graphs, as you like.

OPPENHEIMER—But if you did have on-line capability, you would avoid duplication. I would assume that there is a lot of duplication between your system and the U. S. system.

KOHNKE—I don't know whether it's really a duplication.

NOEL—No, I don't think it would be called a duplication. They have much of the data that we have and vice versa, but it's not an expensive duplication.

OPPENHEIMER—But there is duplication.

NOEL—Yes, but the duplication is in terms of a $12 magnetic tape; that is not a very expensive duplication. The difference is that if they wish to access us across a transatlantic line, that would be rather expensive. The best economics in many cases is putting separate files around the country where they can be used. I imagine ENVIR has some of the EDS data. It's much cheaper to subset at that point and use it a multitude of ways. But I don't think you can really call it duplication in a bad sense.

KOHNKE—The duplication is no more than a magnetic tape.

NOEL—Right.

OPPENHEIMER—Which is $12 a reel.

ROSENFELD—The World Data Centers have an agreement among themselves to do that. That is a deliberate attempt to get certain kinds of data located in certain places.

OPPENHEIMER—But eventually you are going to get to the point where there is so much data that it would be best to have it in one system because then if you have three systems and each one has the capability of using the same data bits, you can compress your system by combining. Then instead of having a tape sitting around which is a duplicate you just have it in one of the three centers and can call for it. And it might be soon that space or capacity may be more important than economics.

BROGDEN—I would like to know, concerning the initial agreement on the format of this sharing of data, what, specifically, you have achieved in this sharing, and how much effort was required to come to this agreement?

KOHNKE—I don't know the answers to all of your questions, Dr. Brogden, but I do know that in fact we are using different formats.

BROGDEN—But the interchange must occur through some sort of a common format.

KOHNKE—Yes, the contents of the differing formats have to be compatible, but then it is only a question of producing the adequate software. The U. S. NODC is familiar with our data system, and we know the NODC system. Simple routines transfer the data from one format to another.

BERG—I think that cost is quite important in this. For example, if you wanted to use a computer from one country to another you have to have either a network or you have to ask somebody to process a job for you, which would incur a delay in obtaining the results.

The only other alternative at the present time is to interchange data and programs by means of magnetic tapes and develop the software to be used on the remote computer for processing. A good example of this is MEDLARS, the medical information system of the National Library of Medicine. There is a MEDLINE (MEDLARS On-Line) network which ties into the TYMSHARE network in Europe so that you can access the MEDLINE network in Europe. There are also MEDLAR centers which can access essentially the same data base or subset of the data base. It has been shown to be less expensive in some cases to maintain a system whose data base is updated periodically, but which is operated locally, than to use a computer network from another country. The local operators were required to use update data on magnetic tape, to create their own software, or to acquire the software and install it on their own computers. So, basically from a cost basis alone, it can be more attractive to pursue the latter approach.

OPPENHEIMER—I have this habit of looking ahead to the next 50 years, and I see the whole world covered with magnetic tape, just like it was more convenient for the United States to develop the automobile and highway communication system than a mass transit system. In about ten years we will be wishing we had a mass transport system and it's going to be terribly costly then to convert. So one has to extrapolate, using all phases of cost analysis, to the future.

BERG—Just take the analogy, though, of paper documents versus microfiche. You have solved the paper document problem and, if microfiche becomes a problem, something different can be tried. You get something even smaller, but it's just a matter of technology.

HELMS—I am certainly very pleased about the direction the discussion is taking now, in particular, because cost is entering the picture. We should all look ahead. I will try tomorrow in the small talk I will give you to give some of the dreams for computer technology in the future. But the fact is that we live in 1974 and there are certain things that we can obtain, but only at a considerable cost. And when we talk about these on-line systems, do we really consider every time if we need the information; the data we can extrapolate. How do you determine what information we need on-line now and here? There is certainly some information where, taking everything into consideration, you will save time and money by using an on-line system. But surely there must be a number of occasions where you can wait or delay until you can have the information in some other form—magnetic tape, microfiche, and so on.

I don't think that we should always compare with the airlines' reservation systems. We tend to do that, but that is one of the on-line systems that really hasn't proven, as far as I can understand, that it is cost effective. Most customers, as well as companies, save considerable time and money in having all relevant information on-line. But really, do we need to have all the data about our environment on line? I think not.

PEACHEY—Mr. Chairman, it's just that I think we can learn a lot from those companies and agencies that have exposed themselves to this very advanced way of doing things. Just think about the airlines for a moment in an operational sense. I am fascinated by the extent to which a developing country can sustain a highly sophisticated terminal to accept a very modern piece of equipment carrying a large variety of people with a whole set of problems and allow them to come in and disperse and re-collect. And I think those kinds of things are going to teach us the points at which it is profitable for scientists to choose highly stylized routines which are sacred and ignored at peril. Elsewhere, for example, in the airport coffee shop there is absolutely no need for any kind of standardization at all except the requirement for payment of services. And I think you see in that kind of world the things that are profitable to computerize, to have rapid access to, and those for which there is no such requirement.

It seems very sad to me that administration and science have not organized themselves in this way. I know the excuses of freedom being interfered with, that there isn't time, and that it would cost more money. But our present methods are a shame on ourselves and our staff in terms of the level of efficiency of work. We simply do not put our own houses in order. It seems to me incredible that you have to take your children to an airport to show them how you can change your mind about a schedule at the last minute and also cancel an alternative booking in another city for an airplane which has been delayed and you can still cancel that reservation because that plane's seating list is still open. I have never been able to take them to any office or laboratory and show them that facility.

LOUDON—I would like to comment on the evils of not having an on-line system. What is happening in our case is that we do a major part of our work through remote job-entry terminals which means a delay of two, three or more hours in getting output back from the computer. This is fine for requirements which can be clearly defined in advance, where one knows more or less what set of data one is going to need in the course of that week and you retrieve it and get it to work on, that's OK, but many users of the system want to interact with the computer processing in some way, and a four-hour turn-around is too slow.

In the course of half a day it is possible to forget what data were retrieved before, so instead of working on one small data set and changing a few parameters and looking at it again, what they do instead is to regenerate the whole job and search the complete data set once more, because they need the ancillary information to remind them of things they knew the day before. Details of earlier steps in a complex analysis are easily forgotten, if the equipment does not respond at the speed of the human mind.

Many people like to work during the day and sleep at night and, therefore, it is convenient for them to do their on-line work during the day. This means that a computer designed for on-line work is perhaps sitting idle for a large part of its time. It seems to me that this situation could lead to transmitting quite a lot of data across the Atlantic. If computers in America were used at a time when Americans were sleeping and Europeans awake, and vice versa, there would be greater usage of the system. I understand that there is an electrical cable joining the supplies in Britain and France because housewives in the two countries like to cook their meals at different times of the day. The total requirement for electricity generating capacity is reduced by passing current through the cable each way. The same may apply in a few years time when we want to make use of excess computer power at a time of day when it would not otherwise be required.

COBB—This is certainly happening now in North America. The Control Data Network switches data between the centers at certain times to take advantage of the time differentials. So I think you have a point there.

NOEL—I would like to say a little bit about a few of the comments. Some of EDS's experiences we were talking about look into the banking and economic world. I think that within the environmental area there is one good example and that's weather information. I'm not an expert in this, so I will just go over it in overview.

The WMO has devised data standards. These data are transferred worldwide every four hours into Washington where they are summarized, put out on maps, and distributed by radio and other modes. I think that there is a characteristic of the data which makes that possible and at the same time economical—it is well defined, or can be well defined, and it is very purposeful. I think the problem that we all have is not with that type of data, but with data that has no well-defined use and for which a system can't be designed as easily.

We should look at other experiences. I think we will find that the data are a different type than the data the economic people are using. One of the problems we have had is that we get data from all over the world, therefore, it comes in different formats. We have bilateral agreements, and for each of these bilateral agreements there is a problem. There is one program that we have to write so we can transfer certain data; it's really not that difficult. But then we find that there are all sorts of other types of data that must be transferred.

The device we have been trying to develop to do this is a data definition form which describes the data in very unique terms, including the scientific meanings of the data. One of the problems that this has brought us is trying to define a rather complete set of vocabulary terms that includes parameters and methods for the header of the file.

We don't know how this is going to work. We have about 2,000 parameters that are coded with definitions. We know what the data means today and as we get new meanings for it we can code them in and keep track of them, they are not just a number any more. Likewise with the methods, we are devising a file so that we can code in what the methods are.

I think this business of transferability really has to be looked at from the point of view of documentation much more than a standard format, per se. We have looked at compressing our data files down in Asheville at the National Climatic Center. We have about 30,000 reels of tape. The NODC station data file is only about 20 or 30 reels of tape. We can compress each of these files by probably about a factor of 100, not a factor of 10. It's quite simple.

The problem is economics again, and dollars do come into play on these files. Economically, it is cheaper for us to store this tape in an expanded format in BCD and not pay the IO cost that it takes to compress it. We know perfectly well how to compress it, but it's not economical. I don't think we can forget the economics of these things, it's very important. Even the U. S. government can't afford to fund us enough. But on the interactive, or the on-line side of it, a number of our files, we feel, are of the type that a person wants to search and then re-do his request on the basis of the information he got out. And this in itself is an economically viable thing. We are trying to put these types of files on-line to make them interactively available, but these are small subsets of large files and again it becomes an economic consideration.

ROSENFELD—If we take this thing to the ridiculous extreme, we could say let's have all the data in the world on-line all the time for whatever reason. That is what this careless discussion is going to lead us to and I couldn't agree more with Dr. Helms. In fact, I am getting conservative. People want to do this or that on-line and want to have a real time system and it is necessary always to ask why. What is so terrible about waiting a few hours or even days for certain kinds of information, particularly if there is a big cost differential? I think we must not lose sight of this, and I hope this Conference ends up as it started, with a diversified approach to a diversified set of problems and with some ability to listen and to understand each other's approach.

I am opposed to the concept of one centrally located data system. The other extreme, with each person working in his own way entirely without communication, is just as bad. We should remember that there are a lot of ways to do things, and a lot of things to be done different ways, and try and achieve a mutual understanding of each other.

EBERHART—Following on that, I agree completely. I guess my own personal objectives lie in two areas for this Conference. My first objective is to get a picture of what kind of information and what kind of information handling systems are available. And the second objective is to establish a common language by which we can talk to each other and exchange this information. So I am looking for a picture of the situation and a language by which I can talk to other people.

PEACHEY—Those of us who are concerned with the emerging European situation, for example, will have, perhaps only with a limited number of items, to do some things in due course in the same way from beginning to end. Already, in the wider scene, substantial progress has been made with harmonized procedures, for example, in the reporting of weather and the occurrence and control of highly infectious diseases. In implementing the Global Environmental Monitoring System (GEMS) we shall be faced at least with agreeing on and developing a new vocabulary.

KOHNKE—I would not agree if you say that in Europe nine countries have to do the same things in actually the same way. I think it is more important that they do something at all, but I would not say that it has necessarily to be done in exactly the same way. I don't like directives to be too strict, because they are not flexible enough to adjust the requirements of a community to new realities.

MAUVAIS—Very often we have this kind of discussion with the Commission and with the other States of the community. I think we must distinguish between two kinds of data. You have data which are scientific data which have to be exchanged between scientists for research, and you have the data which are the data necessary for monitoring of the environment. For the management of the environment very big economical centers are involved. That means when we speak about the economic problems of exchanging data, when we think about the politics of exchanging data we must keep this in mind.

It's clear that when data are necessary for the management of the environment (for instance, data on the health of a population, data on a disease, data on pollution like mercury in the River Rhine) we have too much political and economical pressure on the people in charge of sampling, and on the people in charge of managing the data to reach the possibility of real exchange. If we want to exchange data between, for instance, the Joint Research Center of the Community and the member state research centers for purely scientific purpose, we have no money, no political power to do it. When we speak about standardization and exchange of data we have to establish priorities first in various environmental data and see what we are expecting and for what reasons we want some kind of standardization. I think you really can reach the nine countries of the European Community easier than anyone. If you don't have need and economical power, you will have no agreement.

KOHNKE—I agree, we should make this distinction between pure scientific investigations and the necessity of monitoring our environment to the benefit of mankind.

CUTBILL—I can't resist getting on to the record a definition of compatability as used by Dave Wheeler, a scientist at Cambridge. He defines compatability as a deliberate group policy of not correcting the other man's mistake. And there is a certain amount of truth in that, because it recognizes that in human terms, an operation that works will not be the best possible operation, and our standards should cover what we must do to get something working and that would be within the economic and every other constraint we have around.

PEACHEY—My point was that there are one or two areas where the adoption of common environmental quality standards which affect trade between groups of nations may have a significant mobilizing effect in developing common data procedures in support of those standards.

SCOTT—We were cooperating with our colleagues in Europe long before Britain joined the European Economic Community. The European Invertebrate Survey was started by scientists with common interests in these countries getting together. Now eighteen countries are represented on the Committee. We managed to agree because we were working to a limited aim, namely that we wanted to record what was where at the present time and in the future. We were able to agree on a minimum set of information that everybody was going to record, that the recording methods would be the same, and that this minimum set of information would be exchanged between countries in a standard form. But between those two points each country was free to do exactly what it wanted. If they wanted to record more than the basic information we were asking for, this was up to them. However they processed it, whether they put it into a computer, whether they stored it on punched cards, or just as original documents, was entirely their own business. In this way we are making cooperation work.

OPPENHEIMER—I think it is very useful to evaluate intergovernment cooperative programs to see how the administrators have approached the problem of coordinating the environmental information. However, I must reiterate that there were three premises that we tried to establish for our meeting:

1. That we wouldn't get into an argument about computer hardware, whether IBM is better than UNIVAC.

2. That we would not get into an argument about whose system was better, whether it was 2000 or ENVIR or so on.

3. That we would come out with an impartial recommendation with respect to the world.

Our goal is to come up with a system philosophically that may work. We can use this present discussion on a comparison of different systems to see how different countries approach the problem of data management. I hope that we can develop a unified concept that will allow countries to move toward a goal of compatability. So with that, I think we can close, with your permission, Mr. Chairman.

COBB—Before we close, I would make the suggestion that possibly we can exchange data but we can't exchange information. We might debate that issue tomorrow.

VII. TOWARDS A TOTAL INFORMATION SYSTEM

Session Chairman
Dr. John E. Peachey
Department of Environment
England

Principal Speaker
Dr. Hans Jorgen Helms*
NEUCC
Denmark

OPPENHEIMER—Good morning, I hope the dinner last night rested well with you. It was a pleasure to serve you fish and oysters from our "so-called" polluted bays and to show how rich our effluents really are. Last night I started thinking about yesterday afternoon's conversation, which to me was extremely interesting, and I hope to see it continued today.

There are two points which seem to be arising with clarity. One is that there apparently is no problem with proposed hardware capabilities. The systems that have been demonstrated all have about the same basic characteristics. They may have different aspects in their software capabilities, but in general there are several very good systems already in existence and it looks like data management is mostly a people problem rather than an instrument problem.

As you know, the computer concept evolved from the electronic stage. Electronic expertise created the computer. In order to sell the computers, the developing companies started to build programs for their scientific instruments and I think this has led us into a trap—that we are trying to evolve from a forced system where the companies, in order to sell their machines, have developed their own specialized software in a competitive manner. In many ways we have been a captive of such concepts.

The other thing that comes out loud and clear is the proprietary aspect of data. Here again, we are in a trap. The scientist puts out this material in a publication for which he gets credit, and that's about the only credit that he gets for his efforts. In the university, in industry, and even in government, advances are based upon productivity, with the publications that we turn out being used essentially as a record of our scientific achievements, our prowess, our ability to think, and so on.

---

*Current address—CETIS, Italy

Here is a rather interesting analogy, we publish information from data, if we put the same data in a computer bank it's not considered a publication. And yet, when one looks around the world at the proceedings of many of the marine science institutes, like Woods Hole, Scripps, etc., large monographs appear that are compilations of data. Oceanographic data is collected, printed, and distributed to show what the environment contains. But such compilations are not recognized as a valid journal publication, and are called "grey back" publications. Generally, the scientist gets no credit for such contributions. Therefore, perhaps we need to form our own society, a group, or an ad hoc society in which we can publish a journal which would then serve as the so-called inventory list or the dictionary of environmental data which will comprise a publication.

Most data points collected are paid for with taxes, the public's money. Therefore, such data should belong to the people. And to avoid any proprietary aspect, perhaps we could then publish a small paper describing the data, listing the tape which the data is on, and which bank the data has been submitted to. This identifying characteristic then will maintain itself with the information. So in this sense the people who present information to a computer center would be presenting it in a situation analogous to a publication. And then at the time of advancement when the question is asked, "What have you done last year?" you can say here is my publication list, these are formal publications to learned societies, and this is a publication of my achievements in gathering environmental data. I may not be able to publish an interpretation this year or next year, or even ten years from now because I may be working on a long-range project over a ten - year annual climatic variation. So one would get credit as the data are collected and it would allow the data gatherer to take long-term data and still get annual credit. In the normal journal one wouldn't be able to get credit for the data until the end of the ten years. This latter has generally forced the scientists to conduct short-term projects. So perhaps we are on the verge of some real interesting philosophical changes in environmental data situations.

Another item that has become very apparent is that we can set up a certain list of information which we think is compatible with a perfect environmental data system, like compressability, speed of interchange, maintaining original data, and so on. I hope that this is the type of list that we can come out with tomorrow morning. We should forget about physical and people limitations, and come out with what we think is needed for a total system.

Now I will give the floor to Dr. Peachey for this morning's session.

Peachey—Thank you. Well it looks now as if we are trying to concentrate and formalize some of the key concerns which have emerged as a result of our earlier discussions. It's time to use the ruthlessly clinical approach of Dr. Helms.

In just looking at the list of topics which are relevant to subject interests this morning I think we might, to some extent, add not only "toward a total information system" but also "toward a total information system concept" because if we don't get very far in the vision of a particular approach we shall certainly get somewhere in this emerging concern to see that total systems planning, both at the action level and at the supporting data and information levels, is appropriate to real-life situations and that the thinking of people with direct experience is badly needed to support some of the wider global environmental program initiatives. You can't start until you get a proper disciplined general design.

We did seem to identify that there were points at which you could attach labels to collections of data, some of which corresponded to nothing more than basically a title page for a bibliographic format and others which might concern literally each individual data record. If they did that they would have to take into account data structures within that record, some of which would make accessability more difficult or would require a particular manipulation in order to get to specific elements within the record.

We also moved towards a feeling that it was the cut off point at which you would not even in a technical sense police the intimacy with which you share material, not just for proprietary reasons but because one could not at this stage do the kind of intellectual or systems thinking to see a way clear. That I expect will emerge in our paper and discussions this morning.

In addition, I am reiterating Professor Oppenheimer's point, that we ought to recognize that we do have a facility in the referral system for picking up in the indexing of data files, and this can be done in great detail. That was the whole thinking behind the referral service approach. So we might catch these developments together and exploit them because of their present popularity.

I think if we ask Dr. Helms to put us firmly on the road to disciplined analysis, we shall be in an orderly state.

"Present and Future Computer Capabilities"

Dr. Hans Jorgen Helms

Mr. Chairman, thank you very much for the kind introduction. I hope I don't disappoint you too much by what I have to say because as a matter of fact I hope that some of the details which are on the list will emerge during our discussion this morning and that my job is mainly to provide some background, if I may use that term, and also to try to look into the future.

However, before starting, being the first speaker today, I think that it is my pleasant duty to thank our host and our hostess for the very pleasant evening we had last night. We enjoyed the arrangements here and I will certainly say that I speak on behalf of everybody when I say we feel that we are in Houston; we feel that the attitude of our colleagues here is strongly influenced by the fact that they live their professional life in Houston, or nearby Houston. What do I mean by that? I do not mean so much that the surroundings are pleasant, that we are in a beautiful room, and that we have a nice museum and a nice park outside, etc. No, I am really, of course, thinking and hinting to the technological developments which have been made in Houston or around Houston. The achievements in technology which we all have admired, you know what I am hinting at, as a computer man I must say that some of the computer accomplishments by which I have been most impressed in my life are accomplishments which I see here in Houston. And in saying so I am not thinking about hardware which I saw when I worked here some years ago and was taking a glance through two or three computer installations. Nor am I thinking of the software development which has been done here, although they are also very impressive. What impressed me most in the technological achievements I have seen in Houston are the management skills by which you are able to pull hardware and software together into working systems. Systems which can control some of the most complex situations which mankind ever has learned.

I participated some years ago in a series of NATO Conferences on something we call software techniques, I shall return to that point a little later in my talk. But we did feel at that time that we were, to say in a mild way, rather weak on the management side. I particularly recall a speaker who was responsible for the programming in France in the Apollo Mission, and he really impressed all his colleagues. The reason was that he was able to demonstrate to us that he and his team mastered the management skills. They really mastered the tools they had available. They were able to get the best things out of the hardware even knowing that the programming techniques at that time were and still are, in general ineffective. And that was what really impressed me.

I think that if we review space programs in general that we should not be so impressed by the technology in it and not at all by the hardware in it, but we should be impressed by the way in which people really are able to work together. People with very different backgrounds, with different skills, different talents, (and they cannot all be geniuses, there are so many of them) can work together. But they have been able in the space program, in as much as I can watch it from my computer angle, to work together in a meaningful way. And, as far as I can see, this is one of the biggest lessons for in the world, which I have learned and admired here in Houston.

I do not need to underline once again that I am not a specialist in your subject. I am certainly not an expert on environmental data management, but I am a computing man. At least I describe myself as being a computer man. I am quite humble about being a computing man and I will tell you why, because most computing people who are my age, and this is just around 40 years, cannot really claim that they are computing or computer professionals. I do not think we can claim so. The reason is that we did not have our present subject as part of our formal, professional education and training when we were in our 20's, where we, perhaps, are most adaptive to new things in life. Most computer people, if not all, are my age, some even older, and they have another professional background. Mine happens to be mathematics. Quite a number of my colleagues were engineers by professional background. They have come into the computing field mostly by the hard way and have taken part in what I consider as a dogs race to keep abreast of developments. And certainly we have achieved many things.

There are also things from my own professional life about which I think I have a reason to be rather proud. But all the time I have the feeling that there is a certain solid foundation which I lack. When I look into present computer education as it is taught in our computer or computing science departments at our universities I do not feel so guilty about my own background. Taking a hard look around you will realize that although we are able nowadays to identify a subject which we call computer or computing science, (some call it information science — the name in my opinion doesn't matter so much) the whole field on which our future hopes are based has still a very inadequate foundation, a very, very inadequate foundation where we are striving to understand the fundamentals of what we are doing. Do not expect that any of us will come up with a solution to that in the immediate future. This is of course impossible, a complex scientific field cannot be thoroughly mastered in a decade. It may, in my opinion, take a couple of generations before we are that far.

However, now very great prospects enter the future. Dr. Oppenheimer quite rightly said this morning that he had faith in technology, he had faith in the technological development, and, as I understood Dr. Oppenheimer, also from other comments he has made throughout the Conference, he really has faith in hardware development. He feels that there are large prospects also of interest to us, also of interest for environmental data processing in the trend of hardware development as we can witness them now. There I must admit that I certainly share the faith of Dr. Oppenheimer and I'll give you just a few facts.

Where is technology taking us? This is certainly something you all would like to know. I may be able to speak of some of the directions in which things are moving, but do not expect me or others to give too precise predictions. However, the large companies like IBM, CDC, UNIVAC, etc. often give guidelines on the future technology and in particular on the future hardware. And if I browse back in the literature at the statements which were given years ago and I compare them with what we have witnessed in hardware technology, I must admit that the companies predictions quite often have been correct. Their promises, in my opinion, on the hardware front have by and large been fulfilled. On the other hand, I think that all of you know that the predictions and the optimistic views which these leading companies have taken on software development on applications possibilities have mostly not been filled.

It may be of interest to you to learn that many of us believe one of the most significant current trends today in computing is a strong momentum toward focusing on data bases, the data base of a business and/or of an organization. Years ago when I myself started in computing on a rather big scale, we focused on what we sometimes describe as number crunching. We focused on the high speed of the simple processing unit. We were set up to service physicists, chemists, and crystallographers, who have an immense need for large scale computation.

I have been running a general purpose computing center in my country, Denmark, for research and education purposes since the mid 1960's. The Center where I am can celebrate next year it's tenth anniversary. I have seen a marked trend in recent years in getting away from focusing too much on the number crunching aspect and going more and more toward the data management aspects and data storage aspects.

We are a computing establishment which is available for broad uses in a wide range of scientific disciplines located at various universities and colleges and public research institutes. We have a very different policy, we've been allowed so far to have a very liberal policy about how people want to use us and how much they want to do with us. And this does indicate that we, if I look into our various statistics, reflect the trend in scientific computation nowadays. We used to very strongly distinguish between the so-called scientific computation and so-called administrative computation. In a shop like mine I clearly recognize that the characteristics for administrative computation are becoming more and more progressive and are progressing faster than our characteristics for so-called scientific computation are nowadays. Now this does not mean that we have gone in at our establishments for administrative work like a university's payroll or their inventory, or accounts bookkeeping and what have you, we have a separate machine for these purposes.

Whereas, years ago, just to take one example, physicists in my environment were only interested in having a massive amount of computations done, nowadays I see that they also demonstrate a very clear interest in storing and retrieving data. Had I told my physicists friends that when we started in 1965, they wouldn't have believed me.

Now the problem is, of course, when trends go more and more in the direction of data bases there are at present dreams, if not definite plans, of organizing such a data base as the center of the universe around which an organization functions and to and from which it functions. There are many such statements in the literature and progressive people in the computer industry and amongst computer customers make statements like that. We are moving apparently in that direction.

Then, of course, you can ask the questions, do we have the capabilities to fulfill these promises? And what do we want to do with the data base? Well for one thing, and we have heard it time after time throughout this Conference, we dream of having a data base which is largely if not completely on-line. What do you mean by "having it on-line"? Well in a certain way you can mean that part of the work is performed untouched by human hands. It's not just a matter that you say that you want it on-line because you want to have it rapidly available. It is not all data, in my opinion, which you need to have rapidly available. Within the operations of large scale data bases, there might also be strong operational reasons for having the data base of the organization on-line. You see, in my shop we do not have a very large amount of magnetic tapes around. I think we have a couple of thousand, but of course, compared to what we have heard here at this Conference, a couple of thousand magnetic tapes is nothing. We do know that in an establishment like ours, just to keep up the library and to keep order in the library of a couple of thousand magnetic tapes which are used in the day-to-day operations, is quite a managerial task. It often happens that use of a program calls for tape number so and so. Of course we have a system where this kind of warning comes out in advance so as not to delay operations, but delays may come in that the tape is not available, is not at its proper place, or it might be that in the computer run five minutes before that same tape was used and is now on its way back to its proper place.

There are people, particularly people in business data processing, who also see an operational advantage in having their data bases quite largely on-line. Here, of course, economy, as everywhere, must also be taken into consideration. Is all that possible today? Would all that be a waste of effort today? Certainly, if you think of organizations which might have 100,000 tape reels or something like that around, nobody today in my opinion would dream of putting that on-line. It is possible to do it but economy would forbid it completely.

However, you may hear from industry certain predictions about the development in storage units. And it is really amazing what you hear. I think we at present have a capability on a laboratory basis, not on products for the market, to record on magnetic surfaces at a density of about one million bits to the square inch. Is such laboratory development on the way? We have heard about it, there are publications about it, IBM people, UNIVAC people, and others like to go around to particular customer meetings and tell us things like that. But it has also been predicted that within the next decade it might be possible to routinely use on hundred million bits per square inch. That means increasing today's capability by a factor of 100.

What does all that mean? Well let us go back and take my example of the tape reels. I believe that a 2400 foot tape generally contains about ten megabits of information, that means one hundred million bits. And the predictions are (I haven't yet said anything about timing or about economy) that you can compress one hundred million bits per square inch. It means that the predictions are that in a square inch of a magnetic surface you can store the content of a 2400 foot magnetic tape reel.

These are prospects which might also be interesting for you; these are prospects which might make available some of the aspirations which you have for the future.

Now, of course, if you want to locate this hundred thousand reel library which I mentioned, you will need a mechanism of some sort which can contain this one hundred thousand square inches, on a recording surface. I am not a hardware man and I have not the imagination myself to see how this is possible, but I have heard a number of people from the laboratories of our manufacturers who have claimed that they think they can build a device with this capacity, the capacity in which to use a one hundred thousand tape reel library. Such a device goes into a volume of 3.9 x 5.5 x 5.5 feet, or 1.2 x 1.7 x 1.7 meters. Well, we haven't seen such a box, but in preparation for coming to this meeting I went through a number of papers where people who I generally believe speak with authority illustrate a lot of figures on how the cost per bit stored has gone down all the time. I have such figures I can show you, but I don't think there is any need of doing so because most of you know the details. I have also seen figures on how information gets stored more and more densely.

The origin of these papers (they have been discussed in so-called future requirements projects) is the group of computer users known under the name of SHARE. We have a subsidiary in Europe called SHARE European Association, and there we have been sitting and discussing the future with IBM. Out of all the statements that have been made I think the one which I have selected for you today, and on which I have seen hints from various directions, is this box 1.2 x 1.7 x 1.7 meters in size and in which you have recorded surface of some kind which contains a hundred thousand tape reel library.

But I must also say that when you hear these kinds of predictions, and when you become impressed with them, please also note that I have no authoritative statements on when this sort of device will be available. It is something which they discuss in laboratories. We all know that we are moving in that direction, but we don't know when it will be built; we don't know when it will be available on the market; and, further, we don't know what the cost will be. Our only hope is that the trends, which we have seen for so many years for computer hardware in general, of the cost going down all the time will continue.

And there is a reason for the cost improvement. Why is it possible for the cost on hardware (I underline that I am speaking about the cost of hardware and I am talking about hardware in general) to improve? Well, much of the reason is, of course, that you are able to use a completely new manufacturing procedure, as far as I understand it, when you make these small chips, or whatever, of silicone. Manufacturers are developing more automated methods of compressing more and more electronic capability into smaller and smaller chips. The new manufacturing methods are becoming more and more highly automated, and this is one of the reasons for the decreased costs—the personnel costs can go down in that type of manufacturing. If you go into our present computer manufacturers' plants you will see that although the computers are built by chips and by cards and whatever else, a lot of people are still employed in sticking the units together, and, in particular, in testing them out. Manufacturers predictions again are that manufacturing processes will be so automated that you can avoid a lot of that human element in constructing computers in the future. This should be one of the reasons for the decrease in price.

I can give you just a very, very simple indication of decrease in price in recent years, a thing which always impresses me very much. Not too many years ago, a good desk calculator that could do a number of useful functions cost around $1,000. Nowadays in any bookshop, in any airport lobby, you see these devices selling for around $60 or $70. I always use that as the most impressive, spectacular example of the decrease in price in computer hardware which I know about. Also, the various styles and sizes they come in now—with pocket sizes where they used to be such heavy things that secretaries in an office hardly could carry them and would have to call the men to help.

So you understand from all this that I am really optimistic on the hardware prospects, but I must admit when I have said that, that there are aspects where I am a little less optimistic. I won't discuss the communication aspects. This is very important and there I have also a certain degree of optimism. And I'll tell you why—in telecommunications I have a certain degree of optimism. The telecommunication capabilities, by and large, will be able, in many countries at least, to follow our needs. And that is because our telecommunication people, our telephone and our telegraph people, have been in business for a century. Remember, computer people have only been in business for twenty years. But our telecommunication people have been in business for a century and they are used, in my opinion, to try and predict the future. For generations they have learned at universities how to handle their challenges. And while I know that we can recall many defects in telecommunication, I certainly see in many countries a marked progressiveness in trying to keep abreast of developments. In Europe, in particular, I think I must single out the British Post Office, and my U.K. colleagues may know that we on the continent view with a certain envy the telecommunication plans which we witness in the United Kingdom. There are also other countries who follow well abreast; Germany is one of those which also is very advanced in that aspect.

Can we get rid of the art of programming? Will it be possible for us in the future to put everything into hardware, to build customer-designed computers which did not need to be programmed? There are manufacturers who claim that this can be done to a larger degree than it is done today.

Possibly yes, but we cannot completely get rid of programming. We will have to program; we will have to make software for many years to come. And this is where I do not yet feel there is reason for the same amount of optimism as we have about our hardware developments.

There were two NATO Conferences held in the late 1960's which really had quite a significant influence on my and on many other people's thinking.

Let us very roughly consider the software developments in the past twenty years. You may say that in the 1950's what you really saw as the big software accomplishments at that time were probably the compilers for programming languages. In the 1960's hardware became better, more reliable, and faster, and opened up capabilities which had to be mastered by software complexes known under the name of Operating systems. Throughout the 1960's you saw Operating system after Operating system being built with more and more capabilities, use of facilities, etc. This development came with the development of the large scale time-sharing systems. People were very enthusiastic about it. If you look at statements made at computer conferences in the early 1960's, and perhaps up until the middle of the 1960's, you saw many optimistic statements about what we could do with these glorious software complexes. But as soon as these software complexes became realized in the late 1960's, the disappointment came. The large systems did not fulfill the promises. They did not give us the facilities we had hoped for. The time-sharing systems could not carry the loads which were promised, which they were designed for. Response time was terrible. And above all, cost increased at a rate which nobody had really thought about previously. There was a consensus, not only throughout the computer industry and computer users, but also in the universities, that there was something markedly wrong with our software development. People started getting pessimistic and talked about a software crises. I never believed in that word because the word crises is a misused word in my opinion, and after all, even IBM's OS in spurting versions were able to send a few jobs through computer systems; but, as we all know, not as many jobs as we would have liked within a given time period.

People gathered together, and in 1968 and 1969, we organized conferences in software engineering. I won't go into the details of these conferences. Dr. Rannestad has placed a book here where you can read a very good summary of these conferences, a book called, "NATO and Science" (1967).* The consensus of these conferences was that the so-called mishaps in software, and all the problems we had largely were managerial. They were managerial because people believed that when you had to build a big software complex, like IBM's OS, you just do it by putting more and more people to work, the chinese army approach. But nobody really knew how to have such

---

*Editors Note—This publication has been updated (Rannestad, 1975)

an army of programmers working together. And the appalling results we saw were, in general (and that was the considered opinion of everybody—from academicians, from industry, and from users) that it was largely a managerial problem which we were facing. I won't go into details, because, of course, in software development like elsewhere there are developed quite a number of management tools. But it was realized that it was a managerial problem, as Dr. Oppenheimer said, a "people problem'" which really was the primary problem.

And this problem was stressed in particular in those large firms who otherwise have some of the best management structures we know about, like the large computer firms such as IBM, CDC, and UNIVAC. When it came to software development their methods of management were completely inadequate. And example after example displayed that the best quality of software was that software which were developed outside the large firms.

It is quite normal in research, that the best software are developed by the user groups (the research laboratories, etc.), the reason simply being that they are small. I honestly believe that some of the people there are smarter, on average, than they are in the big production shops for software. But I also think that it is the fact that these groups have been small that is the reason that you see the quality software coming out of them. Similar examples are software houses or university computing center developed software. I think with gratitude everyday of Houston, because we could not function without an adjunct to our operating system, which is IBM's OS, known under the name of HASP. I don't know if everyone knows it, but this was something which was built by a group of smart people here in Houston together with some IBM'ers who were certainly not supposed to do this sort of work. But they simply went ahead in a small group and built it and it's marvelous. It has, of course, since been picked up by the company. I just used that as an example because it was Houston built. I could use dozens and dozens of similar examples and so could my colleagues here.

So this is one "people" aspect related to the production of software. And certainly many of the developments which you hope for in the future will also call for a strong production capability in software, a capability which might not yet be there. Be aware of that. In spite of this fine black storage box I described to you before, which hopefully may be available on a low price basis, there is also (and here I am going back to the data bases) another "people" aspect I see, and that is do you really want centralized data bases? You see I always claim that it is not my obligation as a computer man to tell you what you need, you should tell me what you need and then I will assist if it is reasonable, if it is possible. I can tell you something about the cost, the time scales and so on, but I must say in recalling how enthusiastic users of computers have pressed computer people into many of their developments, that sometimes I think a warning is necessary.

I am like most computer people nowadays, we seem to spend half our time at data base conferences. And, when you go from conference to conference, the problem comes up that there always are social implications of the large scale data bases: Do you really want these integrated data bases? Are you really sure that there is enough political responsibility around to administer such large-scale integrated data bases? I have been somewhat surprised that this aspect has not been brought out yet in our discussions throughout this Conference. I know that you can tell me yes, that there is a tradition one hundred years old of exchanging our oceanographic and environmental data with our colleagues. All right, but there is also a tradition several hundred years old, at least in my country, of keeping church records of people and these data are publicly available. You can go to the minister of the local church and look up in the church book the name of my father, and so on. The problem, as we see it, with the public data bases containing information about human beings is, of course, exactly that this information is centralized and that this information is easily accessible, and at dimensions which are almost impossible to believe. In the future when you have this box I described to you before with one hundred thousand reels of tape in it, these data bases will be even more easily accessible.

I would like very much, if I may deviate somewhat from my assignment, to suggest that you seriously take into consideration that aspect of future developments. I will tell you why I emphasize it. It's not that I want to wash my hands of it, but I am just one amongst hundreds of thousands of people whose life is working with computers. Fortunately, our professional societies (ACM in this country) are becoming somewhat socially/politically concerned about what we are providing our customers. And we do feel a sense of professional responsibility, so at an early stage we will have to tell our users that perhaps we can provide you with all the things you dream of—these massive storage devices on-line, rapid response, fine telecommuniation possibilities and so on—but before these dreams are realized (and the manufacturers will development them), before you sit with these devices and regret that you have put your hundred thousand tape reels into the little black box which can be accessed from anywhere in the world with fantastic speed, do decide whether or not you really want this sort of centralized information about our environment. Do consider the implications of how it can be used and how it may eventually be misused.

That ends my formal message, Mr. Chairman, but I may be able to respond to the more technical questions which we also have on the program this morning. Thank you.

DISCUSSION

PEACHEY—Well, thank you very much Dr. Helms. I think that was a substantial contribution to our meeting. You really gave us some home truths. We are on the edge of a dialogue with an electronic brain and we have to think in that way and we don't know how to do it.

This really brings us back to the problems of a sense of data ownership. Data which once belonged to us cerebrally is passed into an attributable document, or worse still, is merged into an ananymous data mass.

So it looks af if our specifications for total systems thinking will echo both these concerns. The document which has enslaved us into a ritual of publication, refereeing, and storage, (whether that document is a printed piece of paper or basically an electronic document is irrelevant) has also, as it were, protected us from the ultimate exposure of ourselves and our insights and hopefully from unfair manipulations. Our mandate, however, gives us a chance in a sense to accept these concerns and the inherent reservations, but still to press on toward the little black box you spoke of. It is easy to imagine extensions of our cerebral capabilities (as man has always done with the tools he has created), and it would be easier if in some way these little chips and wafers of storage material just happened to be more like our own nervous system. But this is not so, there is the ever-present problem of creating a suitable interface.

After break we might start by considering what the real obstructions are that we will have to accept for the time being. We do have some stepwise approaches which we can identify in a general information system which does not concern individual bits of data but is formalized as with bibliographic services. This country particularly, and some others, have shown that the constraints are very few, that the problems of copyright can be overcome, and that all the insurmountable obstacles can be overcome happily. There must have been a real sense of adventure amongst the people who pioneered this kind of gradual moving together. But bibliographic information had already gone through a kind of ritualized processing with the authors name firmly glued to it and inherent protections.

We might tackle this discussion in a fairly brief fashion by starting with the simplest of possibilities in moving towards a total information systems concept and identifying those points at which it becomes more and more difficult to sustain the concept.

It does seem that during the course of the discussion so far there is general agreement that you can describe and label the whole of what you do for the benefit of other people, whether you are a user, a customer or a potential partner. Many of the conference participants are evidently using this either as a management facility as we mentioned earlier today, or for indexing of data files, or as the first stage in the development of an interdisciplinary program, or for simple housekeeping as in the case perhaps of the Biological Records Center. So we seem to have accepted the fact that the bibliographic document, the published document and even possibly the gray backs, have a role and a reason for them, but that we really do want to move on to something much more flexible. I remember that on an earlier occasion I put it this way, that we were planning these wide initiatives to help rid ourselves of organizational or bibliographic considerations, and I guess that this is implicit in the kind of thinking that we are developing.

Can we learn anything, as Dr. Cutbill mentioned earlier in the discussion yesterday, from the experience of the exchange format for bibliographic material, bearing in mind that this format which was developed for very limited purposes in actual fact ended up being a major tool in the networking of scientific and technical information? If there is anything to be learned there, why don't we have a non-bibliographic equivalent? We need this for inventory taking or referral where we are not referring to a document but to an activity. What futher steps would you need to take in such a format for describing data sets themselves?

Could we start by asking Dr. Berg to just comment on the way in which one might learn from bibliographic exchange procedures.

BERG—Certainly a great deal of effort has gone into the development of bibliographic systems and the techniques by which one can make it easier for someone to get information out of those systems. This is true not only in software, for example, but also in the procedures by which you handle and manage the information in the system. It seems to me very likely that some of those ideas and procedures can be transferred over to this other area. The direct transfer is not reasonable, that's obvious, so one must use the experience gained in bibliographic systems to develop modifications of these procedures and attempt to use some of the knowledge and experience we have gained in established, large-scale systems in order to do this.

I believe very strongly that the significant problems that we have are in the data management area, and in the areas associated with data management. Software is a good example of that, as software is so involved with people and with data management. There are a number of activities that are going on in which there are attempts to use techniques with on-line systems which would help people increase their thinking capabilities. These are very advanced types of research projects. The Advanced Research Projects Agency is involved with many, and other people have been involved with activities in, as I believe someone mentioned yesterday, CAI—Computer Aided Instruction. Computer Aided Instruction has, I think, gotten a bit of a bad name over the last few years, possibly for the reason that many software problems have occurred. Many of the examples of Management Information Systems (MIS) I could give would show that the applications of new technology have failed somewhat. Computer Aided Instruction was supposed to make everything easy, but that has somewhat fallen on its face. Computer aided design itself has not really obtained the potential it could have.

The use of Computer Aided Instruction or MIS techniques could be of great value to users of data systems. Within the computer, the software, and associated procedures, you have ways in which you can output to a person as he is engaged in some kind of activity, with a data system to provide him with instructions or help in the form of a set of procedures as a way to identify and deal with problems as they occur. I think Dr. Peachey mentioned something related to this yesterday when he described how, in a banking system a fellow had a check bounce, as his balance was overdrawn; I don't recall the exact circumstances of the example, but a message could be generated by the computer system to a responsible bank official to watch this fellow, as his pattern of check-writing indicates he may be close to being overdrawn, or whatever the case may be. This is just an example of how systems might incorporate techniques to aid people or assist them as they are interfacing with a portion of the system. And I think that's what can perhaps be done in the environmental data area, but this will be more difficult to work out, as it's not something about which very general rules can be efficiently developed. Perhaps this is possible, and I think that's a good question to ask at this point, is it possible to make general rules, or are these individual cases in which you must do something different for each situation? I do not know the answer to that because I haven't thought about it long enough. Does anyone have any response to that?

LOUDON—Can I suggest that while there is not a set of general rules, because we are dealing with an extremely diverse situation, it does fall into various classes and perhaps we can specify rules that are applicable to some of these classes. Would you mind if I drew you a picture (see Figure 45). I think we can recognize three different extreme situations in this whole area of data coming into the system. First we label this corner of the triangle "data library".

Consider "data library" as being a set of data files produced by different scientists for different purposes. Because these scientists are working as individuals and not together as one integrated group, we find there is a great deal of redundancy between the files. One scientist was in an area to collect information on oil well data so he records their geographic locations. Another scientist looks at the rock types, looks at the paleontology or whatever, and sets up a new data file dealing with the same area. So there is overlap, and the redundant information is repeated in the different files because the objectives are specific to the individual who collected the data.

We can recognize another apex on the diagram, which might be termed a "data base". A deliberate effort is made to eliminate redundant data. Each item of information is stored once and once only, but it is interpreted by different people with different reasons in mind as different people access the same base for different purposes. If information is made available from a data library, the user must accept the data on the terms of the originator, who

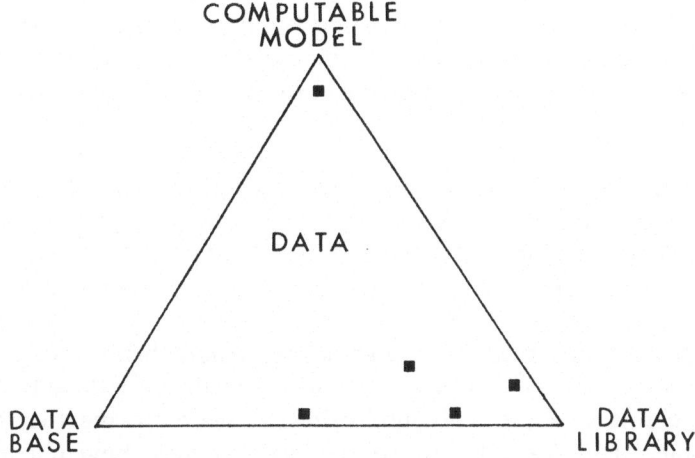

FIGURE 45. The data bank triangle. (A set of environmental data held in the computer can be plotted within the triangle at a position which reflects the degree to which it has the attributes of the three extreme idealized types of data set at the corners of the triangle.)

decided on the objectives and collection techniques which he may have specified in supporting documentation. The user must accept the data on these terms. In the "data base", however, each item of information appears once and once only. If it is relevant to several other items, this is indicated by cross-referencing. One can check for interval consistency between different subsets of the data base. The location of the oil wells in the example must be right for all the applications; cross references to them must tie in. In this case, the onus for correct interpretation is thrown not so much on the user as on the contributor of the data. The contributor must conform to the disciplines of the data base. There must be vocabulary control for instance.

The third apex of the triangle I see as occupied by the "computable model", where one is able to define in advance just what the objectives are, and where one is attempting to predict. A rather general model might be, for instance, to predict the value of a parameter, z, at a point x, y. In a model, one is dealing not only with data but also with relationships between data items. For example, temperature and density of water are related within the ocean. Perhaps the relationships can be stored as a computable model and used, together with known data values, in a prediction at any point x, y of the parameter, z. Like the data base, a large model is somewhat complex and based on cooperation of a number of individuals who are agreed on common aims and objectives. It gives the possibility of evaluating data against hypotheses. The hypothesis, as expressed in terms of a computable model, has the capacity of predicting what the value of a parameter at a particular point will be from the measurements one has up to that time. Thus, one can check new data against the hypothesis, and one can test existing hypotheses against the new data. But this is a rather new area of technology in the environmental sciences, and likely to be relevant to the work of a large organization, rather than an individual research scientist.

The present activities of the organization for which I work are very much concentrated in the "data library" corner of the diagram, with some of our current effort moving more towards the "data base". I would expect more evolution in that direction, and towards the "computable model".

The suggestion which Dr. Oppenheimer made this morning seems to me a very significant one. We should be able to exchange data sets and should be able to regard this as a form of publication. It seems to me that this applies to the data library, and it's difficult to see, for instance, how it could be applied to a large data base where the very extent of the thing makes it difficult to handle in this way. But the data library does follow a pattern very closely analogous to the regular procedure of scientific publication, a procedure which has many attractive feature that we should not lose sight of.

One of these features is that it is a self-driving system. The individuals who are concerned are the authors of a scientific paper, the editor who decides if the paper is accepted by that particular journal, the publisher of the journal, and the reader. Each has motives for carrying out his tasks, and these are built into the system. There is no management problem because the system itself provided the motives for all participants, from the author to the reader. The author is promoted, the editor gains scientific kudos, the publisher makes money (maybe), the reader gains information. The scientific community as a whole benefits from the accumulating knowledge. It also organizes the information on the basis of many individual scientific judgments, whereby references to the work of an author are cited in subsequent papers. If a paper is not felt to be important by the scientific community, it is never quoted, and there is poor access to that paper. At least there was until information systems stirred the sludge from the bottom of the pond. But the scientific literature is constantly evaluated. Good and useful work is accepted, referred to and reviewed. Unimportant and mistaken ideas are neglected, and essentially dropped, by lack of citation, from the information network.

With data libraries there could be an analogous situation, where the "author" of a set of data which are presumably referred to in a conventional paper anyway, is willing to submit the data to an editor of a data library in the same way that he would submit a paper to the editor of a journal. It is the task of an editor to make sure that any data accepted for a library have some relevance to the science and to the library, are original, are not offensive, fit the house style, and are useful, well-presented contributions. If he were in doubt on any of these points I suppose he could refer a data set to a referee just as you would refer conventional papers.

Within a data library there is the possibility of creating a label for each data set, which makes it possible to regard it as a bibliographical item. One could quote the date, the title, the data library and the location of the data set in the library, just as one would give a reference to a conventional paper. I think the concept of a "label" was implicit in Dr. Oppenheimer's suggestion, and was also raised at a meeting of Cogeodata a year or two ago. The "label" means that the data set can be quoted in the bibliographical reference systems, can be referred to in scientific papers, and can appear in the "reference list" of another data set. Again, you have the filtering mechanism, so that data which are important are frequently cited, and are thus readily available to the scientific community; data which are less significant sink to the bottom and are stored forever on magnetic tape, which doesn't cost very much if it isn't used.

So it seems to me that this is an area where progress could be made. I would hope it is not something which requires one monolithic solution. Just as there are various publishers and various journals, I don't see why different data libraries shouldn't concentrate in different areas, with several competing ones perhaps in some areas. This offers an alternative to publishing large amounts of data in the conventional scientific literature; much of it is better communicated in computer-readable form. Put in that context it gives an incentive to scientists to make their data more widely available.

That does not imply that the data are available in the best form for administrators. I think there is a distinction between data for the sake of the science and data which are for management purposes. They are two different situations. Control and management of the environment requires models with a specific, and perfectly valid, purpose and with a scientific input perhaps, but models which are economic and essentially political in their objectives, not scientific. From a scientist's point of view I think it is important to distinguish between these two areas. For the scientist to maintain his integrity as a scientist, his information should be available to the scientific community so that it can be judged by them.

SAYDAM—I should like to draw another picture there which may in a way contrast Dr. Loudon's (Figure 45). I would like to clarify how I understand a total system approach to an information system. In order to do that I will try to put on the board the components of a system related to environmental data (Figure 46). The components that we have been discussing for three days of the computers, let's call them "H" for hardware, "SS" for systems software, "AS" for application software, and let these be one set of a computer. Let "environment" denote another set which we gather our data from. We put the "user" in between, interacting on the one hand with the environment, and on the other hand with computer. This we have been discussing already, and we were trying to figure out how we could easily get the data from the environment into the computer and keep it there in a data base and then use it.

Now let me draw the larger picture (Figure 47) into which Figure 46 fits, because when you talk about a total system you cannot ignore the other elements that interact with total system management. I shall start with the environment first, after all this is the main thing we are concerned with, to collect data from it, how to analyze it, how to control it, how to predict the behaviour of the environment. Let's say that we collect data from the "environment" for our data base. Let me draw a circle here labeled "store", for mass storage, the availability of our system. I shall not discuss now whether we ought to enter on-line or have it there and use it off-line and so forth, I will go into that later on. But why do we store data? So that we can easily access the data for analysis later. So there is another circle called "analyze". But analyzing is not an end in itself; we analyze data so that we can predict the future of the environment and control it. So another circle will be labeled "control". And where does that feedback go? That goes back to the "environment". What we have done, in my modest opinion, although I am not an environmentalist, is place proper emphasis on analyzing, controlling and predictiong, because collecting data and storing it cannot be done effectively without setting your objective.

Now let me draw another circle which encompasses the whole thing, and label it "total system management". What we lacked here was how we are going to manage the whole thing. When you try to manage something you cannot ignore the five basic rules of management, namely, set your objectives, organize, plan, control, measure your results.

How are we going to measure our results here in this system? In my opinion, the better we can predict, the better we can control the environment, the more successful we are. That is the answer, this is the whole strategy from which we can distill our subsets, our objectives. I think this ought to be discussed to a certain degree. I think the analysis of data requires or deserves more attention than how are we going to put data into the computer, and whether we shall have it on-line permanently. I think what we lack here is analyzing.

What I would like to have exchanged is the new techniques of analyzing data so that we can better predict and control the environment. To obtain this objective we refer back, in order to get a better data analysis, to what kind of data, in which order, and so forth I have to put into the computer.

We talk about systems being on-line and so forth and we have been given a lot of demonstrations here, they are all very fine, except I don't think they are cost effective. But you may argue that you cannot see cost effectiveness here where the environment is concerned because the whole purpose of why we control the environment is for our own survival, for our own quality of life. Since you cannot measure quality of life in quantitative terms you cannot really be very cost effective. But our very able speaker this

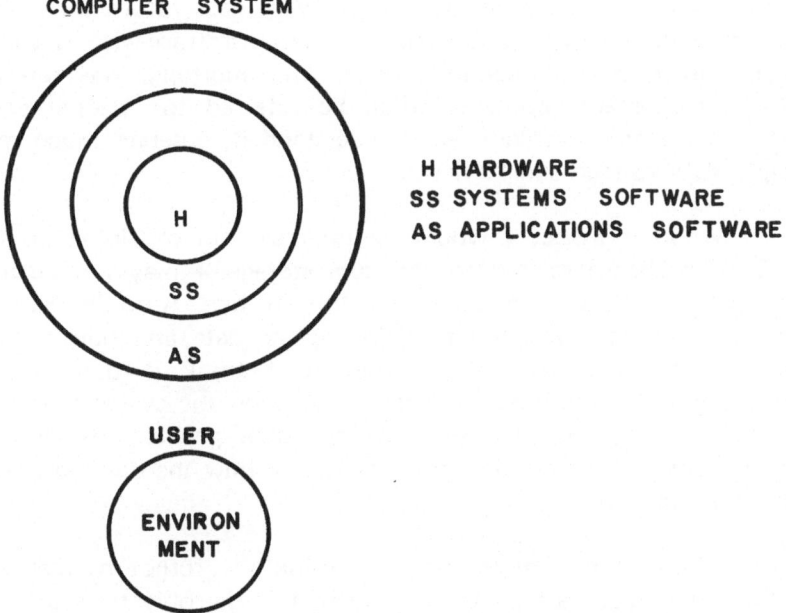

FIGURE 46. Components of a system related to environmental data.

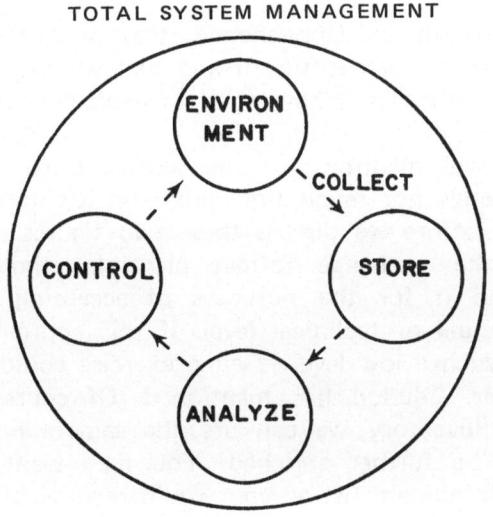

FIGURE 47. Saydam's total system concept.

morning almost assured us that in the very near future we shall hopefully have data base systems cheap. So this problem hardware-wise is solved. I think Dr. Oppenheimer, in his opening remarks this morning, was referring to the real TSM, management problems when he referred to "people problems". And I think this people problem, as I understand it, deserves much more weight than simply data storage and retrieval.

Another aspect I would be interested in obtaining in the future is the software exchange related to the data storage. I may not be interested when I get back to Turkey in biological life in San Antonio Bay, but I may be interested in how people are analyzing the data and how they are controlling data of whichever base they have. So I think if there is an international cooperation, I would like to see it more on the systems cooperation than the data base itself. But if you have a global problem as Dr. Citron has been talking about, then you need not only know how the data base works but also the data that are in it.

There is one more aspect I think we forgot to mention and I would like to draw your attention to, and that is the objectives of such management. I think the objectives should include: obtain and quantify data; store, analyze, publish, and disseminate data; advertize it and market it. So far very little has been said about the advertising and marketing of data even though we have been discussing the fact that even within a country itself we don't know what other people are doing. I think advertising and marketing would reduce redundancies and increase international cooperation in this respect.

Again, I agree with Dr. Oppenheimer that present hardware possibilities and availabilities are more than sufficient and that we should give our attention to the other three features that I have tried to explain in this verbage.

PEACHEY—I will mention, for the record, some of the earlier points because obviously there's not much time left, and it's important to insure that these are highlighted before we discuss these two figures. We did identify that we could borrow the exchange format principle (though not its present structure) and extend it for the purposes of accepting material from other people at the intellectual or technical level. If this approach is widely held to be acceptable, then such a low level labeling exercise could be applied to all of the programs that Mr. Loudon has mentioned. Of course from the point of view of the general inventory we can use the same kind of labeling facility which may need to be further enriched. You may want to list, as it were, records of computable models just as you would records of data sets or of data bases.

The next point that was made was that Computer Aided Instruction is like everything else, it's had a bad time. Nevertheless, CAI provides an opportunity for people to reduce their fear of these new approaches. We should not forget that one of the major mistakes that was made in exposing people to newly computerized bibliographic systems was that on-line techniques for perfecting questions came in far too late, not for getting the answers but for doing the equivalent of browsing.

Another point is that data managerial attitudes vary greatly according to one's work situation. Then there was an observation from Dr. Helms which is relevant to these considerations and we ought to highlight it too, that sadly the administrators have made more progress than the scientists. I think you have to accept that the people who look after payrolls think just as much of integrity as a scientist publishing his paper. You know they have their ink wells and pens and they were just as deeply upset as we would be. Therefore, I guess we are going to have to make major professional concessions in surrendering the ritual and self-driving procedure that leads to conventional publication.

Now having caught these observations for the record, and I know there are lots to come, I will ask Professor Oppenheimer to comment.

OPPENHEIMER—I think your comments, Dr. Peachey, are extremely relevant, that we are now apparently on a threshold of changing the whole philosophy of data management, and I speak not from a manager's point of view, but from the users. As you said, we have to look at the problem of data management from both sides. Dr. Saydam approached it from a manager's point of view, it's quite different when one looks at it from a users side.

The simplicity of Dr. Loudon's statement attracts me because this is the type of simplicity that might be workable when the concept is put on paper and offered to the scientific community. Does the system give us the validity that we hope to achieve as contributors to a data base?

Dr. Saydam's diagram is also relevant because it's simple and it's workable. However, the concepts he gave are somewhat different when one looks at it from the environmental side. First of all, the cost effectiveness that he mentioned is relevant. We are spending billions of dollars on wrong decisions on our environment today, and these decisions do affect our economy. One of the reasons we are having an energy crunch and financial problems today is because of the forcing action of pollution control. Our automobiles cost us $700 more this year because of the pollution control devices. Cost effectiveness, as Dr. Saydam very aptly pointed out, must be considered because humans are a part of the environment.

The other point is that while Figure 47 is useful to incorporate the total data system, there are several ways in which data can be incorporated in a computer. The National Science Foundation's GEOSECS program is a good example to use here. The Naional Science Foundation's IDOE (the International Decade of Ocean Exploration) and the International Biological Program are an attempt to evaluate our environment at this period in history so that we can record changes that occur in the future, base our predictions upon them, and then extrapolate changes that occurred in the past.

In contrast, the system shown by Dr. Toledo M. to describe the Flora of Veracruz is an inventory system that had an objective that is not equivalent to the mathematician trying to solve a mathematical problem or chemical equation. It's an attempt to get some record of the environment, and will have many, many inputs of different types of information.

SAYDAM—Dr. Oppenheimer, may I introduce another idea which I didn't mention, that system as small as it is cannot be identified by itself alone. This dynamic model (Figure 47), must be incorporated into the other system parameters such as gross national product and so forth. Therefore, as you say, cost effectiveness and the results of this system within the broader systems dynamics, or world dynamics, is quite relevant.

OPPENHEIMER—The ecologist who is trying to get people to pool their information so that we can make better management decisions, is trying to increase the cost effectiveness, so we cannot disregard cost effectiveness. Therefore, we don't have to justify our data system by saying that we can meet its cost because the justification is already evident. We must go to the computer type of system to understand our environment and, therefore, our environmental control costs can be used to justify the system-type approach. How can we get people to cooperate in a system for pooling information so that the data base can be pooled to make decisions?

PEACHEY—Before I go to the next speaker, can I go back to Dr. Helms' contribution. He warned us that we have this terrible burden of software and there are many of us who think that models are the next in the series of these burdens between data and access.

Could we now hear from Dr. Rannestad and then Dr. Citron and Dr. Rosenfeld.

RANNESTAD—We collect data from the environment; we put it into storage, either as a data file or library, or as a data base; we analyze the data; make predictions; and, in the end, want to control the environment, as indicated by Dr. Saydam (see Figure 47). This is the scientist's way of working. If you are a manager or a politician, however, you go the opposite way. You start with the decision you want to make, or ask the question you need an answer to. To answer the questions, you may have to prepare a program to analyze the data. If you do not have the right data available, you may have to collect the date, put it into storage, analyze it and then to back to the control of the environment. It is time-consuming to collect data, and it is, therefore, necessary that we have available data in such a form that it may be used not only for the original reason for which it was collected, but also for additional purposes. We should try to find some way to make this possible. That is, data banks must be constructed and described in such a way that they are interoperable.

ROSENFELD—Dr. Oppenheimer seemed to be emphasizing some concern about proprietary data. I don't think that that is a major issue. There are certainly those who are really concerned about the proprietary nature of scientific data, but it's usually only for a year or so. Also, the amount is not very great relative to the mass of data that are being collected by organizations and people who really collect them with the viewpoint of public use. I think at the moment the proprietary aspect is a kind of red herring.

PEACHEY—I think that is a very useful thing to balance against the more general point that was made.

KOHNKE—I am not quite happy with the system identified in Figure 47. I entirely agree with Dr. Oppenheimer that this has been constructed more from the point of view of a data manager or data producer and not so much from the data user's point of view. I think it does not really reflect the critical point in the whole system. Therefore, I would like to draw up another scheme which will give some modification (Figure 48). First, of course, there is the environment which has to be observed and investigated. But most of the data will be collected for quite specific purposes. Thus, what really happens is that each data set will be analyzed under different special aspects without taking into consideration the data or the results of colleagues who are working on a bordering area of the environment. This is what I call the critical point in the present data systems, because what is needed for a better understanding of the environmental processes is the input of data from different disciplines. And to me the arrangements for interdisciplinary investigations are still very insufficient. What may be the reasons for this? I believe one explanation is the (historical) situation we are living in; it is a transition phase of technology and of the human generation. It is easier for young scientists to get familiar with modern computer techniques. Another point is that with disciplines gathering more qualitative data, the inclination to use computers for data processing is not nearly the same as with scientists handling digital values. And finally, not yet all scientists realize that the pure description of an environmental state does not tell anything about its evolution and how it will change in future. This can only be solved by interdisciplinary work.

PEACHEY—Thank you very much. I just don't know how anybody is going to describe all this in words, but it does truly reflect the range of interests. I was very interested that when this was taken up by one previous contributor I made a note that the difference was that administrators start with the action that seems right on the top and then generally verify, but a scientist starts with the observation and modeling. What we are getting is a reflection of the different points where people start from and we don't yet know how to connect these.

But now I have time to call on Dr. Cobb and then Dr. Cutbill.

COBB—I shall be brief. I just wanted to make a comment expanding our discussions this morning. The information on Figure 48 is little more than a change in the presentation of our previous discussions, but by putting it in this format I think that it relates closely with the questions of data exchange we raised yesterday.

Basically we have a number of levels of management in any organization where the highest level defines and sets up "goals," the middle level in management will set a series of "targets" to reach a specific goal, and at the bottom level a series of "objectives are set in order to reach those targets (Figure 49). Now, within a single discipline I think that we are capable of providing good goals, targets and objectives, we are able to collect our data, and produce the analyses and results we are interested in. You can come in at various levels here either at the "control" level or at the "data" level as we said this morning. But once we start to get into multidiscipline areas as we do with the total environment picture, it is very difficult for us to define our goals, targets and objectives in a very specific manner.

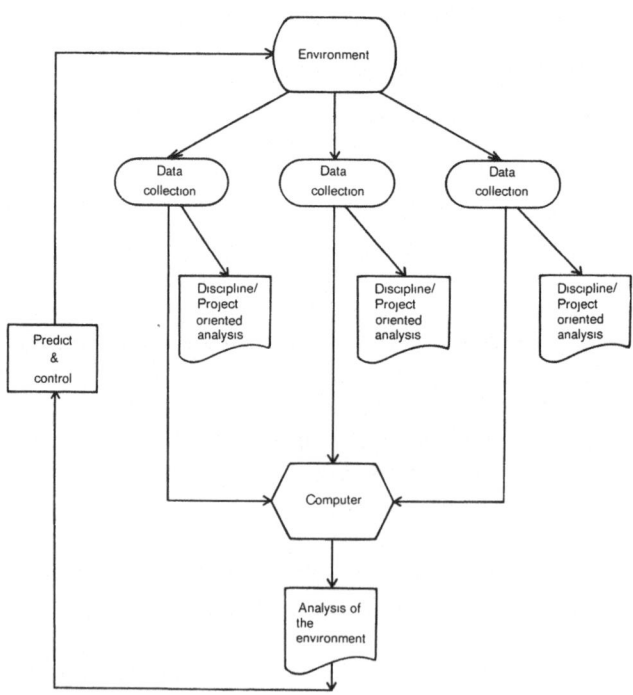

FIGURE 48.   Kohnke's modification of Saydam's total system
             concept (Figure 47)

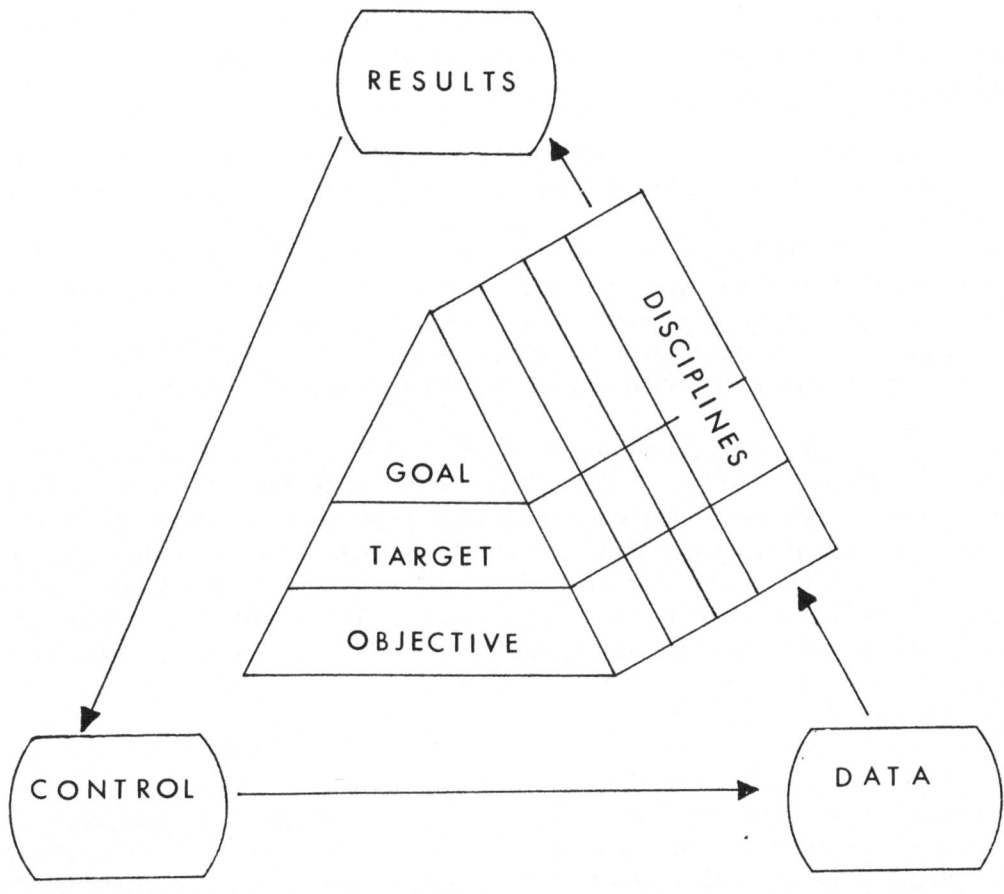

FIGURE 49. Flow pattern for data management goals.

When we start to consider data management systems, which I think is really what we are here to do, we have a problem because we don't know what the data is that we are trying to manage. I would be interested in other people's comments on this. I feel we can perform the required functions of data management and exchange within a discipline, and perhaps do it within one or two disciplines, but once we go across the complete spectrum of all the disciplines within the environment we are unable to define the problem in detail, let alone resolve the problem.

PEACHEY—Well, thank you very much. I think that kind of situation leads to the setting up of special cross-sectoral surveys and appraisals, which sadly, are often not firmly integrated with on-going routine data-gathering, which thus are left untransformed and the special surveys often re-collect the relevant data all over again.

I would just like to ask Dr. Cutbill to make a contribution, mainly because in the biological field he and his friends have done some really original thinking on formats for individual records.

CUTBILL—I have been reluctant to get technical, but there are problems here that are technical problems, that are, I think, worth looking at. There's no doubt at all that whatever comes out of this Conference people are going to go on communicating data. There is equally no doubt at all that our data communication will be very limited because we need a common bank and it has been quite clear that even with the English-speaking people, English is thoroughly unsatisfactory for our present purpose and scope of data. We are not communicating that well today, or for the past two or three days.

At all levels of communications you have to get your language terms right to be successful, for instance the example that John Peachey has been pushing on us of the bibliographic exchange. They were successful and in fact created a common language that has very little to do with computers at all — it does have a few things because it is stored on magnetic tape. Basically, they could agree on a language for describing the information they wanted to communicate, but librarians are generally very narrow-minded people. They wanted the author, the title, the book size, the number of pages and how much it cost, and they were successful.

Our analogy to that I suspect is the number of bits per inch and the number of tracks on the tape, and so on, and no doubt if we got down to that it could be done, but it's a nuisance, it's not a problem. Is there anything we could learn from that about communicating our affairs scientifically? At first sight there isn't because you have done a sort of token job, and the moment you start talking about your data this is not quite in your field. Apart from all the concern about total systems, actual communication is always in a much smaller area with many more motivations and limits. We have tackled this problem in the museum field concerned with a range of descriptive data. It's about the same sort of scale for any kind of raw data system. We have found that within that sort of area you can find enough common ground to build yourself a language in which the structure is defined and some of the terms are defined. Of course, not all of them are defined and some are acquired as you go along. You achieve a level of description that is adequate for you to communicate and you need to establish inventions for passing on the inventions.

You end up being able to get a record that is self-contained, that carries its complete description of how you put it together, the antecedents of the data and so on in terms that all the others will understand and disentangle. It doesn't mean the data's compatible. It doesn't mean it's usable; but at least you have got the information in a state in which the receiver can make judgment. You have got it across.

I have no particular desire to go into detail and we don't have the time anyway. The technique is useful at all levels and it's useful even to be thinking about this. It leads to something which I think John Peachey feels very strongly about and that is the size of the unit of information that you can devour. Now, I am not sure what angle he comes at it from, I suspect it's asking questions to keep the minister off balance in Parliament. I come at it as a geologist and yet I am not a working geologist. My starting point always is quite small pieces of information that I want to pull out of some system. I don't know how library people approach their communication problem, but they must package data into great big packages and I then have to go through a quite complex treatment to get it back out. In our application large groups of data are re-sorted down to the small pieces I want; and then I find it has all gotten stored the wrong way. The sort of descriptive technique that we have used does give you a chance to move your information around in much smaller packages, the size of unit that you actually want.

Thus, there are a series of techniques in museum collections that people should look at for their own requirements. It doesn't matter which computer system they are using, there is this descriptive technique that is worth much more, and I think it should be on the record.

PEACHEY—Thank you very much. I think that is a very effective summary you gave and I think we should move on to the next session.

VIII. TOWARDS A TOTAL SYSTEM

Session Chairman
Mr. Dieter Kohnke
Deutsches Hydrographisches
Institut, Germany

Principal Speaker
Dr. William Brogden
The University of Texas, U.S.A.
also
Dr. Bruce Lighthart
Environmental Protection Agency
U.S.A.

KOHNKE—In the past sessions of this Conference, we have heard a lot about scientific, sociological, and political necessities for an improved exchange (availability) of information and data. Dr. Helms has given us an idea of the technical possibilities that will enable the ever-increasing flow of data and information to be made transparent and available to everybody.

In the course of the Conference, we were made acquainted with the already existing information systems (ENVIR) and the arrangements for the exchange of data (International Oceanographic Data Exchange—IODE; World Weather Watch). Though these systems are already working very successfully, they are still too specialized, i.e., they always refer only to one small section of the total environment.

In our final session, we are now going to try to develop a concept for a general environmental data management system. The objective to be achieved by such a system is quite obvious. It shall help scientists to get a better understanding of the causal relationships and assist politicians in their planning and decisions. The data can be fed into models that bring the physical, chemical, and biological components into a functional connection by means of which the environment—after a sufficient verification of the model through the environment itself—can finally be monitored and its behaviour predicted.

Meteorology is the most advanced discipline in this respect, I believe, where a worldwide agreement could be reached on a system of observation and exchange. This finally made it possible to reproduce the behaviour of the atmosphere in numerical models and to predict it. Through the establishment of the Global Telecommunication System (GTS) the availability of meteorological observations is now outstanding.

Dr. Lighthart has agreed to present a quite different model this afternoon. This is the model with which his laboratory center in Southeastern Montana tries to assess the impact of man's activities on a terrestrial ecosystem. Here they try to approximate, by means of a model, changes of the environment (with regard to animals and plants) that are caused by coal mining and by heating power plants with coal.

First, Dr. Brogden will give us some results of the so-called "systems approach" to ecology and try, in a kind of similarity study, to apply H. Odum's idea of a "special energy circuit language" to the complex processes in environmental data management.

I think that Dr. Lighthart should speak immediately after Dr. Brogden, before we start with the discussion on the necessity and the form of an environmental data management system. May I now ask Dr. Brogden to begin.

"The Possibility of Unifying Concepts Which Could Be Used In The Evolution Of A General Information System for Environmental Science"

Dr. William Brogden

The most notable trend in the environmental sciences recently has been the trend to the "Systems Approach" to the environment. I would like to look first at some results of the systems approach to ecology. Later I would also like to consider the earth sciences and environmental management. It seems to me that although the systems approach has had some important influence on data gathering efforts, we have not made sufficient use of it in data handling as yet.

I have been struck by the thought that it would be interesting to take a systems approach to our subject here by inserting into the "World Dynamics" model a component called "Environmental Information," which would both consume resources and partially regulate and modify other components of the model. It would certainly be hard to get the right coefficients, but it might help us to evaluate the importance of what we are actually doing.

The steps involved in the application of the systems approach, according to its advocates, are the following:

1. Decide on the system boundaries.

2. Determine what you are going to treat as separate objects within the system.

3. Determine the behavioral features of the objects.

4.   Determine the interactions which occur:

    A.   between objects within the model, and

    B.   between objects and things external to the system.

Some of our graphic attempts on the board earlier were attempts at system definition, and I don't think the systems analysis people would be completely happy with them because I am not sure that we have defined our objects or the behavioral features of them too well. To relate this to data handling, I think that in the systems I am familiar with any rate, our data systems are frequently able to describe the objects within a system, but do not do well at describing interactions. This is partly because we do not measure the interactions, and partly because we do not have a good data structure to represent the interactions.

In an example from my own experience, an estuary can be defined by system boundaries pretty nicely if it has a defined shore line. Your sampling programs can tell you the quantity and diversity of fish, what concentrations of chemicals are present, etc. But it is much harder, both to determine and to represent in a data structure, exactly what the fish are doing, or where the chemicals are going.

Recently an environmental scientist, Kenneth Watt, has taken the systems approach, and presented a set of fourteen "core" principles of environmental science (Watt, 1973). I would like to discuss several of these principles which I think may shed some light on the problems of environmental data management. The first two are simple restatements of the first and second laws of thermodynamics.

1.   All energy entering an organism, population, or ecosystem can be accounted for as energy which is stored or leaves the system. Energy can be transformed from one form to another, but it cannot disappear or be destroyed, or be created.

2.   No energy conversion system is ever completely efficient.

One of the consequences of these principles is that we can use energy as a basis for modeling and describing environmental systems; I would like to spend a little time considering bioenergetics models later.

The third principles has no corresponding law of thermodynamics or physics.

3.   Matter, energy, space, time, and diversity are all categories of resources. (Watt defines a resource as anything needed by an organism, population, or ecosystem which, by its increasing availability up to an optimal

or sufficient level, allows an increasing rate of energy conversion.

Now, we are used to thinking of matter and energy as resources, but if it is true that the other three are of equal importance, I think this will have some consequences for environmental data management. An example of space as a resource might be the space through which a population of organisms is spread, if there is too much space between individuals, there is an increased chance of individuals finding a space, if there is too little space, there may be behavioral disturbances. Another example of space as a resource might be the interstitial spaces of a sediment which are available for colonization by microorganisms.

Time is a resource. An example might be the length of the growing season at a high latitude, or the numbers of years between destructive hurricanes on the Texas coast.

For an example of diversity as a resource, consider the difference between a predator which can feed on only a single food species, in comparison to one which can feed on a wide variety of food species. The predator with a single species of prey will be vulnerable to anything that happens to that species, while the other will be affected very little if a single species of prey is no longer available. Of course there are many successful organisms that are specialists, so this principle is a generalization which may not be true in specific cases. Diversity within a system comes about both through the number of objects within a system and through the number of connections.

The implications of this principle for data management are that a "total information system" would have to be able to represent and keep track of all of these resources. At the present time, environmental data systems seem to concentrate on keeping track of matter, space, and time. This is probably due to the fact that these resources are by far the most frequently measured, and the conceptual schemes for storing and retrieving this type of data are fairly simple. Energy will not be too difficult to handle under present schemes either. However, I think that representation of diversity is going to be somewhat more difficult, particularly when you consider that when we start storing data about an organism or an ecosystem, diversity will be poorly defined. Only during the course of years of investigation will the interrelationships become clarified and perhaps altered by additional data. This means that our data system will have to be exceptionally flexible in the representation of interrelations.

Figure 50 shows two ways of representing a food web, both directed graph and undirected. This is the representation of diversity both of object and interconnections within the system. The study of this ecosystem might have started out with just information about the abundance of these different organisms, later investigations might have established major interconnections as shown. Next year new organisms or new interconnections might have to be added. It will certainly take a very flexible system to accommodate this growing amount of data.

A fourth principle which Watt (1973) has proposed is that:

4.   The steady-state diversity of communities is higher in predictable environments.

This illustrates requirements for our data management system, we are going to have to have methods for evaluating the predictability of an environment. This can be approached statistically, of course. It might be as simple as evaluating the standard deviation of salinity within a certain area of an estuary, or it might take more complex statistics.

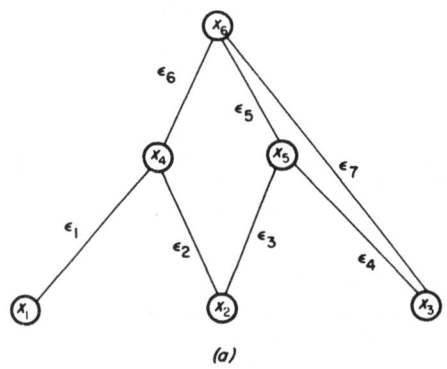

(a)

Example (a) undirected and
        (b) directed graphs.

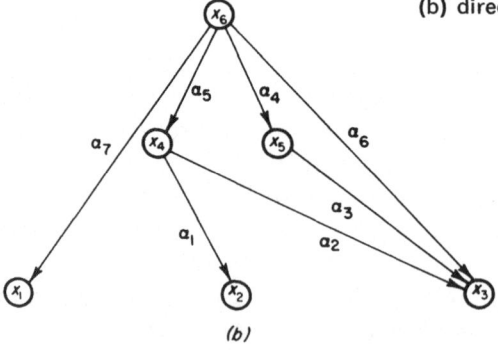

(b)

FIGURE 50.   Representation of a food web as a directed and as an undirected graph. (Taken from Gallopin, G. C., 1972. Structural Properties of Good Webs, p. 241-181. In B. C. Patten (ed.), System Analysis and Simulation in Ecology V2, Academic Press, New York.)

Two more principles related to this one are:

5.   The ratio of biomass to productivity increases through time in a stable environment, up to an asymptote.

6.   The diversity of any community is proportional to the biomass divided by the productivity.

We could spend quite a long time on the discussion of these principles and the remainder of the fourteen, but I would like at this time to move on to the discussion of the use of energy for the description of ecosystems as exemplified in the work of Howard T. Odum (1971).

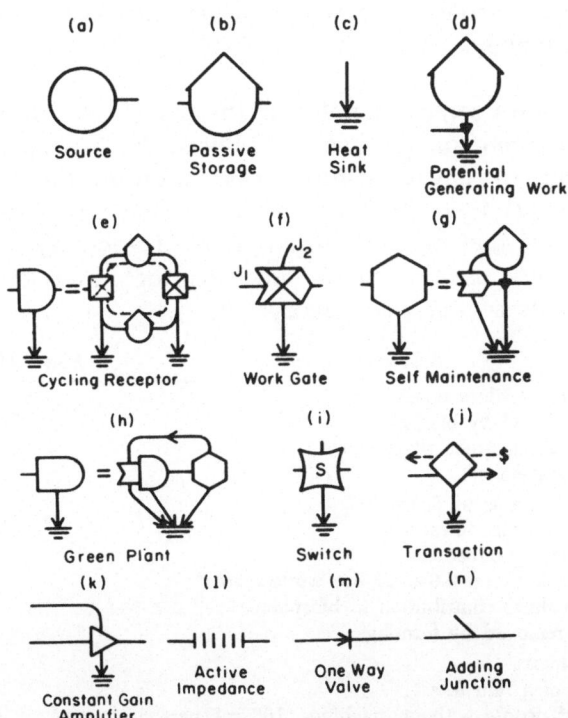

FIGURE 51. Symbols utilized in Odum's energy circuit language. (Taken from Odum, H. T., 1971. Environment, Power and Society, p. 34. Wiley-Interscience, New York. 331 pp.)

Odum makes use of a special "energy circuit language" to describe systems, the individual circuit elements are shown in Figure 51. The "energy source" (a), and "passive storage" (b) symbols are straight-forward, the "heat sink" symbol (c) shows the energy loss required in an energy transformation system by the second law of thermodynamics. Symbol (d) is a module which represents the storage of potential energy. Symbol (e), is a "cycling receptor" module which might represent an energy receptor such as chlorophyll which is excited by adsorption of light energy, then passes the energy on and returns to a lower energy state. One of the important properties of such an element is that the rate at which energy can be absorbed depends on the availability of receptors in the low energy state. The "work gate" (f) is a unit in which the flow of one energy makes possible the flow of a second energy. The "self-maintaining module" (g), which might correspond to an organism, is one in which stored energy is fed back to do work on the processing of more energy. The "green plant" (h) module combines the energy receptor with the self-maintaining module. The "switch" unit (i) is used for the control of flows which are either on or off. The "transaction symbol" (j) is used for the description of human economic systems, which not only involve the flow of energy but also of money.

Figure 52 shows some examples of the magnitude of energy flows to be found in the environment on the basis of kilocalories per square meter per day. This is an example of environmental information in the form of power or energy flow. Energy flow can be used to express such things as the consumption of fossil fuel or the consumption of food by an oyster reef in a Texas bay. This gives us a common language for interrelating a tremendous variety of components of the world ecosystem.

System	Power (kcal/m²/day)
Incoming energy (dilute type)	
Sunlight absorbed by biosphere	5110[a]
Sunlight reaching green plant level	3400[a]
Maximum conversion[b]	170
Organic production as in Figure 1–5	
Production of a rain forest	131[c]
World primary production	6[d]
World agriculture on 8.9% of the world's area[e]	0.26
World agriculture contribution to biosphere[e]	0.024
Production removed by farming[f]	3.6
Consumer systems	
Respiration of a rain forest	131[c]
A village of people without machines, 100 m²/person	30
Fossil-fuel consumption in the whole biosphere[g]	0.135
Total consumption of biosphere (production and fossil fuel)	6.1
Fossil-fuel consumption in the United States (per U.S. area)	3
Animal city, Texas oyster reef[h]	57
Fossil-fuel consumption in a large city[i]	4000

FIGURE 52.   Examples of the magnitude of energy flows found in the environment in units of kilocalories per square meter per day. (Taken from Odum, H. T., 1971. Environment, Power and Society, p. 50. Wiley-Interscience, New York, 331 pp.)

Now, before looking at some example systems described in this language, I would like to say that I am not trying to sell the idea of converting all environmental data to an energy basis. What I am trying to say is that energy does provide a useful basis for the intercomparison of widely differing systems, especially those including man's activities, and that, therefore, a "total data system" might be expected to be able to manipulate some things in terms of energy.

Figure 53 shows a very simple representation of the world as a system in terms of three sectors — agriculture, industry, and government — showing both the flow of energy and the flow of money. Actually, this circuit has been too simplified because the input of fossil fuel has been ignored.

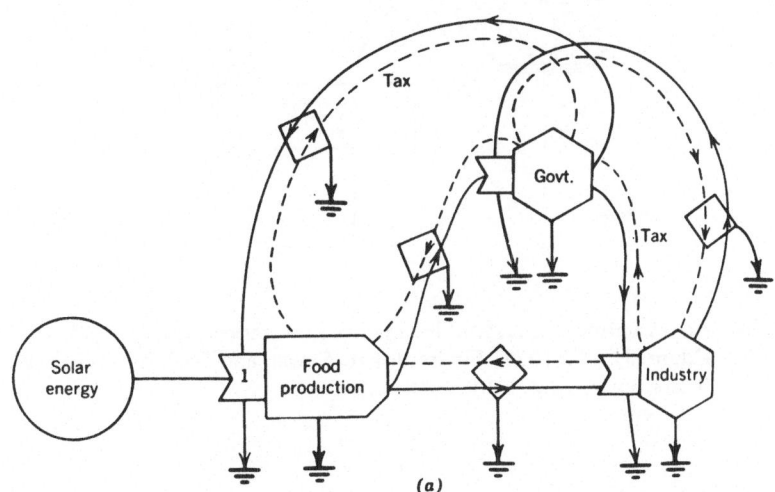

FIGURE 53.   A simple representation of the world showing both the flow of energy and the flow of money. (Taken from Odum, H. T., 1971. Environment, Power and Society, p. 181. Wiley-Interscience, New York, 331 pp.)

Figure 54 is an attempt by Dr. Odum (1971) to represent the interaction of an estuarine ecosystem with human economy and it involves both the energy flows within the ecosystem and the energy flow within man's economy. The estuary is the Corpus Christi Bay system, which is about 200 miles southwest of Houston. Only a few numbers appear on the energy flows because many of them are very hard to evaluate. The use by industry and the work done managing the ecosystem are evaluated by an equivalence between energy and the known monetary flows.

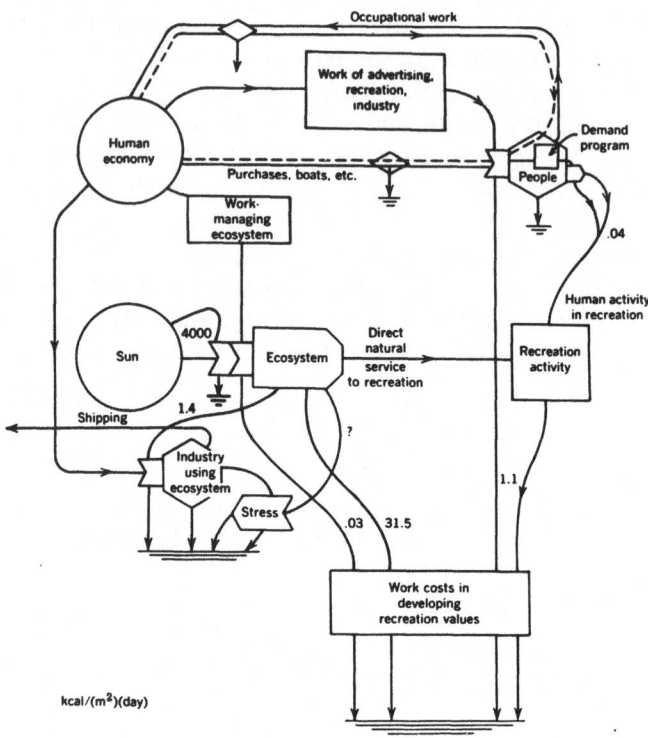

FIGURE 54. An estuarine ecosystem in energy flow representation. (Taken from Odum, H. T., 1971. Environment, Power and Society, p. 297. Wiley-Interscience, New York, 331 pp.)

I calculated that the activities of research projects I have been involved with recently which are devoted to understanding the ecosystem of this Bay represent an energy expenditure of about 0.002 kilocalories per square meter per day. In this ecosystem at least, man's activities represent only a small fraction of the natural energy flow.

With that I would like to leave the consideration of bioenergetics for now, and consider the earth sciences for a few moments.

In geology and geochemistry, I believe there will not be too much trouble in getting wide agreement on terminology, and the requirements for a "total information system." For inorganic geochemistry, at least the properties of the elements are pretty well known.

Geochemists divide the elements into these main subdivisions:

1.   "Atmophiles" which tend to end up in the atmosphere,

2.   "Lithophiles" which tend to concentrate in the crustal material,

3.   "Siderophiles" concentrating with iron and nickel in the earth's core, and

4.   "Chalcophiles" which have a strong affinity for sulfur.

In addition to these groupings, we of course have the similarities which the periodic table leads us to expect, such as the similar behavior of zinc, cadmium, and mercury which are of recent concern.

The dynamic transformations of these elements are generalized into the geochemical cycle, here shown in abstract form (Figure 55). The large part of this cycle runs very slowly, and the dynamics of it are not likely to be contained in a "total information system". The dynamics of the smaller cycle within the dashed line are certainly of interest. It is within this part of the cycle that interaction with the biosphere occurs.

ROCK CYCLE AS A DYNAMIC SYSTEM

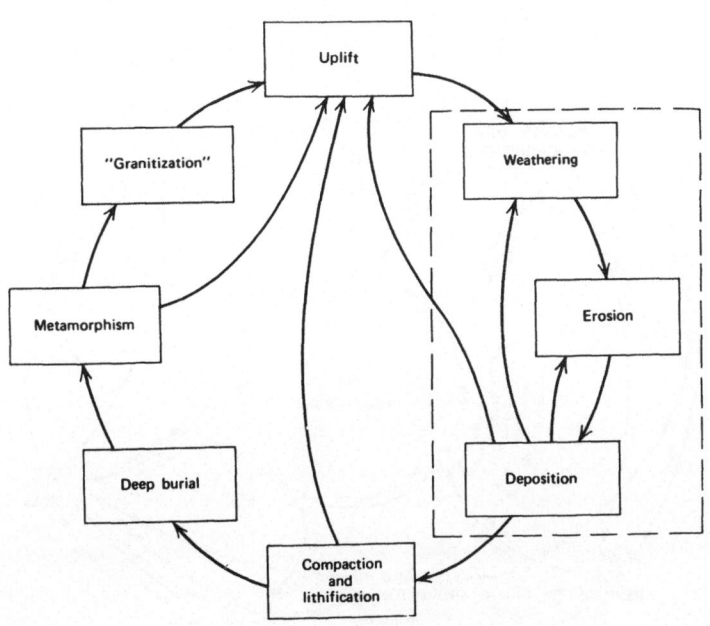

FIGURE 55. The geological cycle of elements. [Interaction with with the biosphere occurs within the dashed line.] (Taken from Harbaugh, J. W., and G. Bonham-Carter, 1970. Computer Simulation in Geology, p. 21. Wiley-Interscience, New York, 575 pp.)

Let us suppose that I as a geochemist decided to study the phosphorus cycle in a particular area which might be represented by Figure 56. We see the tremendous number of interactions with organisms, with soil, within fresh water and seawater.

The natural geographical unit for organization of geochemical information about the elements would seem to be the watershed, and I think this is going to be one of the geographical tools which we would want to have for a "total information system." Watershed boundaries and flow relationships can usually be defined and agreed on, and I do not think we will have much difficulty in describing geographical locations in terms of watersheds. In fact, the STORET system operated by the Environmental Protection Agency has very good representations of relationships within the watersheds. Certainly the information on phosphorus in fresh waters would be found to be organized by watershed; however, information on man's activities might be found to be organized on another geographical basis such as by county.

This just serves to illustrate the fact that a "total information system" would have to be able to inter-convert between several different geographical bases.

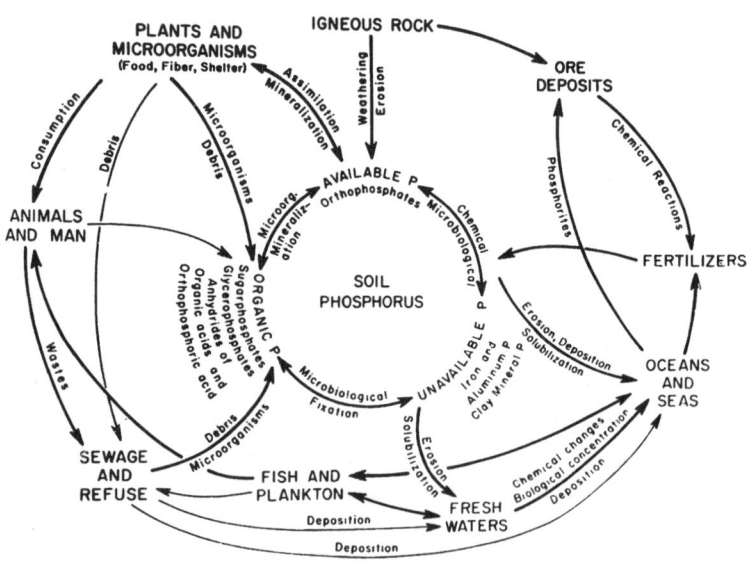

FIGURE 56. The phosphorus cycle. (Taken from Fairbridge, R. W. (ed.), 1972. The Encyclopedia of Geochemistry and Environmental Sciences, p. 951. Van Nostrand Reinhold Co., New York, 1321 pp.)

Other geological fields are very heavy oriented towards the use of maps which represent painstaking interpretation of surface samples, well cores, aerial photographs, etc., and the use of computers for the production of these maps is already substantial. The problem of graphic representation of variables in three dimensions also occurs in oceanography and meteorology, even though the time scale for describing the dynamics of the systems ranges from thousands of years to hours. It seems likely that a "total system" could share a great deal of software between all of these applications.

Geological data, however, requires the representation of discontinuities in three dimensions. Sometimes these are sharp discontinuities between rock type, or more gradual changes such as soils. Also, the precision with which the features are located must frequently be much better than that which suffices for oceanography or meteorology.

One of the recent important trends in the United States has been the requirement for the production of environmental impact statements. This has caused a tremendous change in the environmental sciences in this country and has brought a whole new wave of demand for environmental data. The U.S. Geological Survey tried to produce a guide to the evaluation of environmental impact (Leopold, et. al., 1971) and I think it's interesting to look at what they considered important.

Incidentally, there are a number of other experimental approaches to environmental impact evaluation, based on some sort of matrix crossing man's activities with their effect on the environment.

In the USGS process (Leopold, et. al., 1971), environmental information enters the stream of decision making with the characterization of presently existing conditions. There are stages of identifying impacts of the proposed project on the environment as it exists, followed by assessment of these impacts by means of a matrix. Alternatives are also evaluated through the same system, and finally a recommendation results. I think that one advantage of a "total system" would be that everyone concerned with evaluation of the environmental impact of a project could have access to the data describing the present condition of the environment, even if they could not agree on the significance of the data.

I think that this is the most important aspect of the social responsibility of environmental data management, that it is dangerous to have only a single segment of society that has access to data and, therefore, can say that it knows better than the rest of society which directions to take because only it has command of all the data. It should be the responsibility of environmental data people to make sure that everybody has access to this data which represents, in this case, the condition of the environment and which is used in evaluating environmental impact and policy decisions.

That is all I have to say, thank you ladies and gentlemen.

KOHNKE—I thank you very much. Before starting the discussion of Dr. Brogden's paper, I would like to ask Dr. Lighthart to tell us something about his work. After his presentation, we shall discuss both lectures.

"Computer Modeling Techniques For Assessing
The Environmental Impact Of Strip Mining On Terrestrial Ecosystems "

Dr. Bruce Lighthart

In coming to this Conference on data management, I found myself rather a "duck out of water" because I am not a data manager; I am a working level biologist. My Director had asked me to talk about the kinds of data manipulations that we are doing in our research laboratory, but I think within the discussions that have taken place so far the kinds of data manipulation we have been carrying on have largely been mentioned and it would be redundant for me to talk about them any further. Contemplating that I might contribute to this Conference, it occurred to me that I would like to leave you with the notion that there has been a lot of oceanographic and atmospheric data collected, but little terrestrial data collected and that copious amounts of it will have to be gathered and managed if we are to use the predictive tools for environmental management that I am about to talk about.

Scientists at the National Ecological Research Laboratory (NERL) are carrying out experiments in the Colstrip area of Southern Montana (Figure 57) in order to assess the impact on terrestrial ecosystems of strip mining of coal and its use in power plants.

The potential magnitude of this problem is indicated by observing the pink colored strata (Figure 58) that appears just below the surface rocks in the canyon layers. The pink color is the result of the underlying coal burning and baking the overlying clay mineral a pink color. The overlying clay subsequently fell into the steam left by the burning coal, smothering the fire. Thus, the red baked rock gives a visual impression of the extent of the coal strata formation.

The extent to which biological processes may be interrupted by strip mining activities such as shown in Figure 59 are indicated in Figures 60 and 61, where large amounts of forest vegetation could be destroyed, and cattle feeding on grass whose growth could be inhibited might at best be temporarily stopped.

Mined coal is to be used in coal-fired power plants that are being constructed (Figure 61). The smoke stack at the Colstrip plant is 500 feet (less

National Environmental Research Center • Corvallis
(Complex Includes NERC Headquarters, 9 Associate Laboratories, 5 Field Stations)

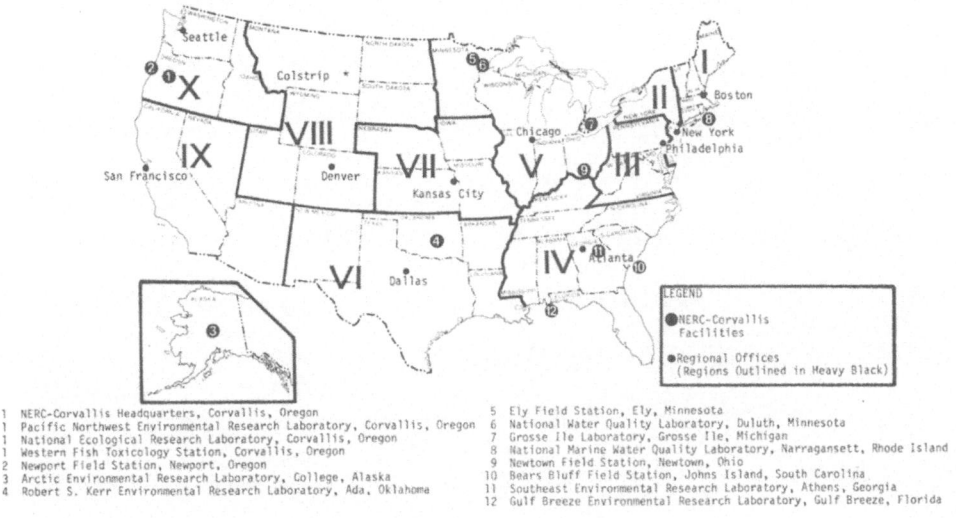

1 NERC-Corvallis Headquarters, Corvallis, Oregon
1 Pacific Northwest Environmental Research Laboratory, Corvallis, Oregon
1 National Ecological Research Laboratory, Corvallis, Oregon
1 Western Fish Toxicology Station, Corvallis, Oregon
2 Newport Field Station, Newport, Oregon
3 Arctic Environmental Research Laboratory, College, Alaska
4 Robert S. Kerr Environmental Research Laboratory, Ada, Oklahoma

5 Ely Field Station, Ely, Minnesota
6 National Water Quality Laboratory, Duluth, Minnesota
7 Grosse Ile Laboratory, Grosse Ile, Michigan
8 National Marine Water Quality Laboratory, Narragansett, Rhode Island
9 Newtown Field Station, Newtown, Ohio
10 Bears Bluff Field Station, Johns Island, South Carolina
11 Southeast Environmental Research Laboratory, Athens, Georgia
12 Gulf Breeze Environmental Research Laboratory, Gulf Breeze, Florida

FIGURE 57.   Map showing NERL study area where grassland simulation model is being modified
and adapted in an effort to predict the effects of strip mining and coal-fired power
plant impacts on the local biota.

than 200 meters) high.   This is considered to be a small plant for this area.
Thus, beyond the strip mining effects on the biota, one might further expect
that the gasses emitted by the power plants will also contribute a significant
deleterious impact on the local biota.

The mining process is shown in progress in Figure 59 and the
"reclaimed land" after cessation of mining in Figure 63. You will note the
poor outgrowth of vegetation in the mined-over area of Figure 62.

OPENHEIMER—What percentage of the total acreage per year is
affected?

LIGHTHART—I am sorry, I don't have that kind of figure offhand.
Right now, it is relatively small, but the number of energy companies that are
going into that area and their plans for the future will markedly change this
ratio. We are talking about hundreds of square miles. It is unbelievable when
you are there and you see the pink ridges, ridge after ridge of them. These
ridges extend as far as you can see from an airplane flying at 3,000 feet above
the ground. So, it is potentially a huge problem.

FIGURE 58. Aerial photograph showing fired pink clay overlying a coal seam in the NERL
study area (see between arrows).

One of the ways we are trying to get an understanding of what is going to happen to the ecosystem that we will be impacting is to prepare predictive models to test possible consequences of the mining activities on the local biota. One such model has already been developed through the U.S. IBP program for a grassland environment. As you saw in the figures, there was a lot of grassland in the Colstrip area. NERL and Grassland Biome scientists are adapting and modifying the grassland biome model to the Colstrip site for subsequent use as a predictive management tool for this complex system.

In an overall view, the model is constructed in several interconnected submodels: an abiotic module regulating temperature and water relationships; a soil nutrient module for phosphorus and nitrogen cycling; a plant module, including morphological and phenological characteristics of several groups or species of plants, 15 or less consumer animal species or groups, and a decomposer module.

My aim in presenting the following detailed information is to give you some idea of the complexity of the model system we are talking about and the copious amounts of data that we are going to have to manage in order to simulate one particular kind of environmental system into a model. For a more detailed explanation of the grassland simulation model to follow, see Anway, et. al (1972).

The model is essentially a point source system, i.e., an integrated one square meter of land. The abiotic portion of the model provides a stochastic

FIGURE 59. Aerial photograph of active strip mine in Southeastern Montana.

FIGURE 60. Aerial photograph showing forest lands.

source of water in the form of rain water. There are mechanisms in the model to move the water through the soil and from the measurements made in the field, fairly good predictions are made by the model. There are also mechanisms that transfer heat down through the soil. Both water and temperature effect the biological processes going on in the soil. Coefficients incorporated into the model and their subsequent verifications, are being done by automated instruments so there is a copious amount of data generated.

The nutrient portion of the model is made up of a number of field measurements transformed into rates of transport of nitrogen and phosphorus species in the soil to-and-from state variables. These rates are affected by temperature, moisture, and biological processes.

FIGURE 61.   Aerial photograph showing forest lands and cattle activity (trails generated by cattle over a two-hour period in freshly fallen snow can be seen).

The model has the potential for storing and manipulating a large number of species and their sub-anatomical pieces, for example, a warm- or cold-season plant broken up into above- and below-ground, alive and dead parts. Transfer of materials between plant compartments is also included in the model. For example, the shoots may be eaten by a cow at different rates. Also, in the model is an information tabulation of the nitrogen phosphorus content of each one of these pieces of plant material. Further, some of the pieces of plant material will die and go to the decomposer section to be decomposed; others go to the consumers. The plants respire, that is they give off CO_2 which is eventually transferred to an atmospheric carbon pool.

The consumer compartment in the model may be only cows, but can be up to 15 other kinds of animal consumers, both predators and herbivores. The food preference of the animals are included in the model, which means that a lot of data has to be collected because the preference of different kinds of plants is different for different animals and for different times of the year. Animal food preferences may be obtained by putting the test animals on the range and allowing them to eat naturally. By having a fistula in the side of an animal, ingested food can be removed from the animal's stomach and analyzed to evaluate preferences.

FIGURE 62. Small coal-fired power plant at Colstrip, Montana. (Stack is 500 feet high.)

The last portion of the model that I will talk about is the decomposer module: bacteria and other microorganisms in the soil. Decomposer organism respiration is affected by temperature and soil moisture as well as the inorganic nutrients, nitrogen and phosphorus. Products of decomposition are recycled to nutrient pools for subsequent uptake by the plant consuming portion of the model.

Many of the measurements necessary to "force" the model are made by automatic and tape recorded instruments contained in a trailer like the one shown in Figure 64. The trailer contains various analytical instruments, including gas chromatographs and atmospheric monitoring devices for gaseous and weather measurements (Figures 65 and 66). All of this information is digitized, put onto magnetic tapes, and run back to the laboratory for analysis.

FIGURE 63.   Aerial photograph of "reclaimed" land at Colstrip, Montana.

I hope that I have given an impression that this is an extensive model and that it will require a large amount of data input if our experiment to use this kind of system is to be able to predict the effects of man on ecosystems. Finally, this is essentially a single-point model, i.e., not taking into account biological gradients; therefore, one might anticipate that the foregoing model will be only one element in an ultimate array of interconnecting components in a gradient model.

FIGURE 64. Air monitoring trailer.

FIGURE 65. Automated chemical analyzing machines.

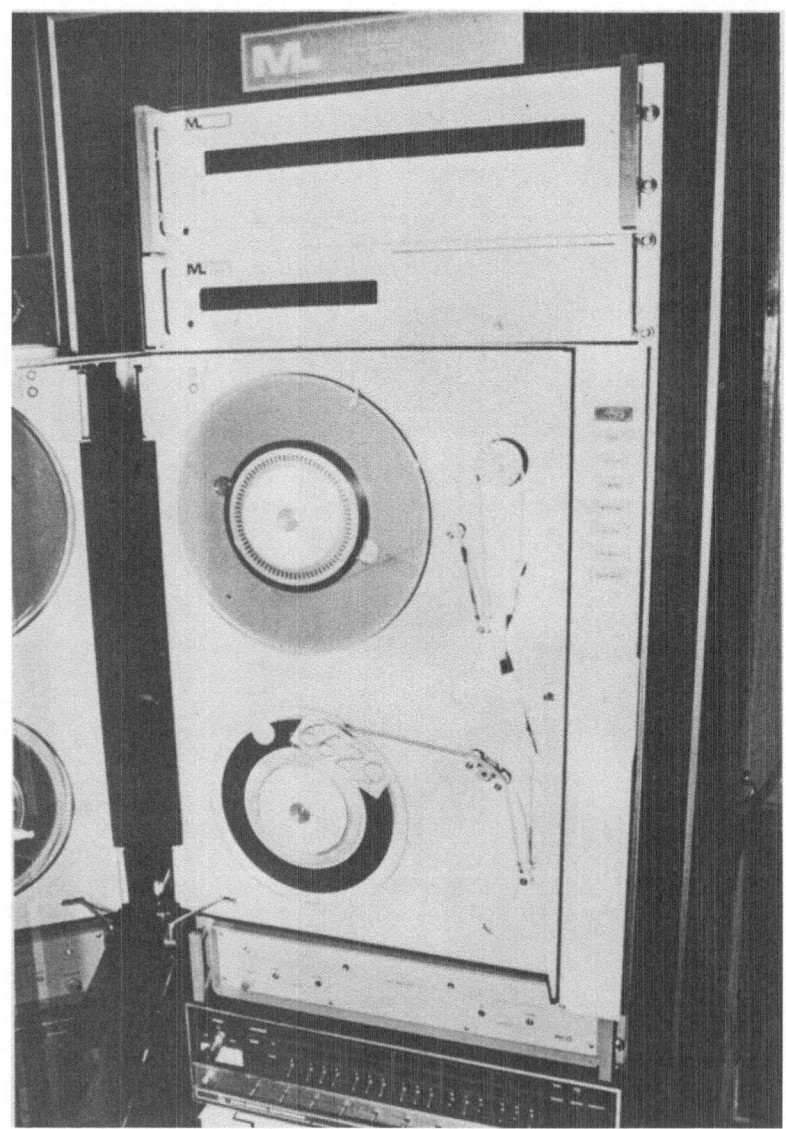

FIGURE 66. Computer and tape recorder for chemical analyses measure at the Colstrip, Montana site.

DISCUSSION

KOHNKE—I want to thank both Dr. Brogden and Dr. Lighthart for their very interesting and instructive lectures. It was particularly on account of the contradictory views they contained that I found these two lectures so stimulating — On the one hand (Brogden) the attempt was made to make the environment more transparent by abstraction to the essentials; on the other hand (Lighthart) there was the demonstration of a practical, and, I think, successful example to recognize the interaction of the individual environmental factors in a limited and more or less closed system.

I think that we should discuss in the subsequent debate the question of how far the use of Odum's "energy circuit language" — perhaps in a modified form — would promote the conception of an environmental data system. Furthermore, I feel that we should bring about a connection between the theoretical explanations and the practical implementation of an environmental data management.

There is another point which became quite apparent to me after Dr. Brogden's lecture. There seem to be discernible evolution stages of the various scientific disciplines. What I mean is, and it was surprising to me, that biologists do not look so much for interrelations and interactions between chemistry, physics and biology, if I correctly understood. The interest of biologists may be of quite a different nature from that of physicists or chemists working also in the marine field.

That may mean that the requirements for the secondary data users are not of a uniform character for the various disciplines. In marine physics, and for the most part, also in chemistry, I think most scientists left the stage of merely describing a prevailing situation. They left that stage because they could no longer compile really new and directive knowledge. They have extended their research to investigating the interactions between the subsystems (atmosphere, hydrosphere, biosphere, etc.). One example is the investigation of the upwelling phenomenon in the oceans. It cannot be explained without knowing the meteorological conditions. If you are trying to understand the ecology in such an area, you must take into consideration things like the ocean's dynamics, the supply of light, nutrients, and plankton, just to mention some important factors. What I am trying to say is that a better knowledge of the environment as a whole can only be achieved through an intensified interdisciplinary research. And this implied a wide-range environmental referral and data system.

OPPENHEIMER—May I interrupt? I think that you have made a very interesting point. The examples that Brogden and others used can fortify the concept. The whole purpose of a unified data system is that each scientist must place his information in a general system. It is exactly as you say, the environment is not composed only of data from the physicist who is interested in the salinity and chemistry, and perhaps wave forces and dynamics of water movement. However, if he places information into the system and, at the same time, there is a biological expedition taking the same information but adding biology, material from the physical oceanographer and from the biological oceanographer provide supportive data and the pieces start to fit together.

We must not confuse the fact that many of the items we have discussed today are the interpretation of data and not the handling of data. Once we have the data in a usable base, then the physical oceanographer and the biologist can interface and say really we were pretty close together three years ago. The data base for salinity over the Atlantic was within one month of your biological program; perhaps we can see some similarities and results. Until we, as ecologists, reach the point where all the data sets go into one common system we can talk from now until forever and still have problems in relating the large amount of biological data being collected in various programs throughout the world. We must speak in a common data language. We must use valid and common terms whether they be salinity, nitrogen, meteorological symbols, energy, the distribution of a life form, or what cows eat. These items can be put into a common language just like the 25 toxicants that are to be used in the Global Environmental Monitoring System. We aren't talking about interpretation, we are talking about plugging that data in so that after four, five, six years we can start to see trends.

The data base in Texas now will show relationships between environmental data. We have five years of salinity data for some of our Texas bays; we are learning things about dynamics of our bay systems we could never find before. Because somebody took a measurement, and somebody took the time to put that measurement in a data system, we can go back and ask for that piece of data and relate it to all the other data. If we don't like the data we can go to the source and ask who recorded it; what went wrong; is it a real anomoly or is it bad data? The person who originally collected the data can interpret it relative to his needs, as long as he sticks that data in a library so others can use it. Then the data becomes more meaningful.

Until such time, we just have to divorce the proprietary part of it, which has been automatic in the past. We must acknowledge the general need for collective environmental data that prodes this meeting.

LIGHTHART—I just want to make one thing perfectly clear. I did not mean to imply in any way that the model that I was talking about does not require all sorts of physical and chemical data. It absolutely does. Just one example; you can't make the plants grow without sunshine and it has to be rather precisely defined as a driving force in the model. So we have to have all this physical and chemical data. As a matter of fact, I might say that the biological systems are integrating all of these physical and chemical things going on on the planet and without it you don't have all the little pieces that you need to grow an organism.

KOHNKE—Then you are studying interactions?

LIGHTHART—Yes, absolutely.

BROGDEN—I didn't want to say that we didn't study the interactions, I wanted to say that we didn't do too well at representing the interactions in the data form, if I can make that distinction. I don't know if I can or not. It's a problem of representation. In other words, if one is measuring a rate of ammonia consumption by a certain organism, it is a more complicated thing to describe than just a concentration of ammonia.

EBERHART—It's hard to measure that particular phenomenon. It's harder to measure rates than to measure volume.

BROGDEN—In other words we are doing pretty well in describing so much ammonia in a soil concentration. It requires a small number of descriptors to peg that measurement down. But it requires a more complex description scheme to represent the things that you measure in the way of interaction. You have to include both the fact that you took soil, nitrogen and maybe the bacteria; and you measured a rate. Representing the interactions is more difficult than representing the concentrations of the object of the system.

OPPENHEIMER—The more data you get into the system, the better chance you have of using the data for environmental interpretation.

BROGDEN—OK, but I'm just pointing out the problems in the biological.

RANNESTAD—I think we are talking about two different systems (see Figure 67). One is the conventional system where the data you want to analyze is your own, or data you have received from other scientists and put into your data bank, as discussed by Dr. Kohnke. In this case you know the validity of the data and it is, therefore, possible to have faith in your analysis.

On the other hand, when we are talking about a large central data bank, or possibly several data banks for use in our analyses, you no longer handle the data personally; and, therefore, don't know its quality. To be able to use the large data banks and have faith in your analyses, you must be sure that the data has a minimum quality and is in a form that you can use. It is, therefore, necessary that all data be checked by an editor before it enters the data bank. Both types of data banks should be included in a future system.

ROSENFELD—I do want to thank our gracious host and hostess for the fine time we are having here; and therefore I feel a little embarrassed by having to challenge directly what Dr. Oppenheimer just said. And if that was in the record I have to take the opposite view. He said, I believe, that anyone can

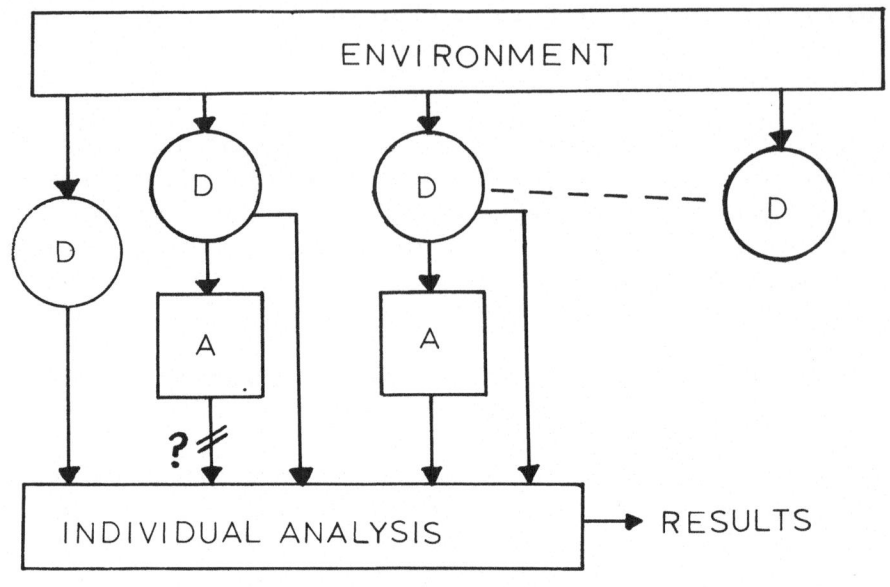

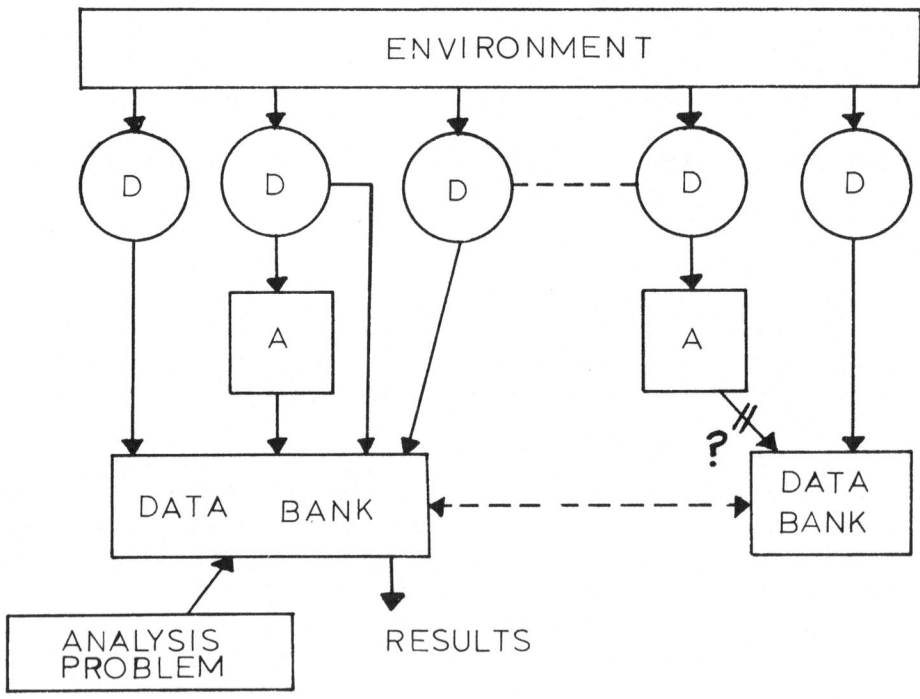

FIGURE 67.   A comparison of data systems. [In the first graph (a conventional system) the
                     validity of the data is known, in the second graph (a large data bank) the validity
                     of the data is not known.]

interpret data. I feel that I would tend to put it around the other way that anyone can collect data and very few can interpret it.

The last talk by Dr. Lighthart is, I am quite certain, what Dr. Loudon was referring to yesterday as the model for collecting the data and I hope that this example must lead us now to realize that there will be no kind of a single approach or single system. I cannot visualize the data which are collected in an experiment of this kind being in a data bank from which anyone else can obtain those data and interpret them in the light of the kind of model Dr. Lighthart was talking about. I can see the reverse. I can see that at the beginning of such an experiment it would be very nice if he had a data base of, for example the weather and so forth in Montana at that location prior to starting such an experiment.

So the diagram that Dr. Loudon put up this morning is the kind of thing that I hope this Conference will lead to. I personally am a firm believer in the integrity of local data bases managed in a way (which relates to the lower left-hand corner of Figure 45) relative to the experiment for which they were collected, maintained perhaps by that organization. I don't think any data should be lost. I think we all agree that somewhere around the world we should maintain data. The meteorological data, the massive observations by satellite, and much of the oceanographic data that are taken either as part of a designed experiment, or routinely, have a great deal of value, even apart from the particular designed experiment, and must be contrasted with the kind of data that are taken in the experiment we just heard about.

I would hope that in the next half day or so we can look toward at least a dual kind of system where the data library be treated, as several people have suggested, as retrievable on some kind of a bibliographic data base. This would be somewhat parallel to a bibliographic data base operated and managed by the people who collected and used it for a specific model. It should maintain its integrity based on the model. And again there is the other kind of a generalized data base. If we can aim toward that kind of dual approach perhaps that might be what we are here for.

MARGALEF—I would like to draw a sketch to try to relate different levels of conceptualization as I see them. As an ecologist, I am strongly discipline-oriented; and, moreover, I am not familiar with computer techniques. We have here in Level I (see Figure 68), a number of boxes representing different variables, as salinity, temperature, results of some chemical analysis, fluorescence of water, number of particles as the output of a Coulter Counter, and so on. Behind this is some preliminary model about definition of the environment, availability of methods and capability of doing the sampling. These variables are operationally defined and I see that this is what goes into the bank of data. The organization of the bank of data, it seems to me, cannot be completely static, but must allow some changes for the better. If some variables are highly correlated, you get redundancy. Some of them can be dropped; other variables can be added. Technical improvements can be introduced that may change the definition of the variable; and perhaps the stores can be saved and made compatible using some conversion factor. Perhaps such capacity for internal adjustment has to be considered as a matter of course in every bank of data.

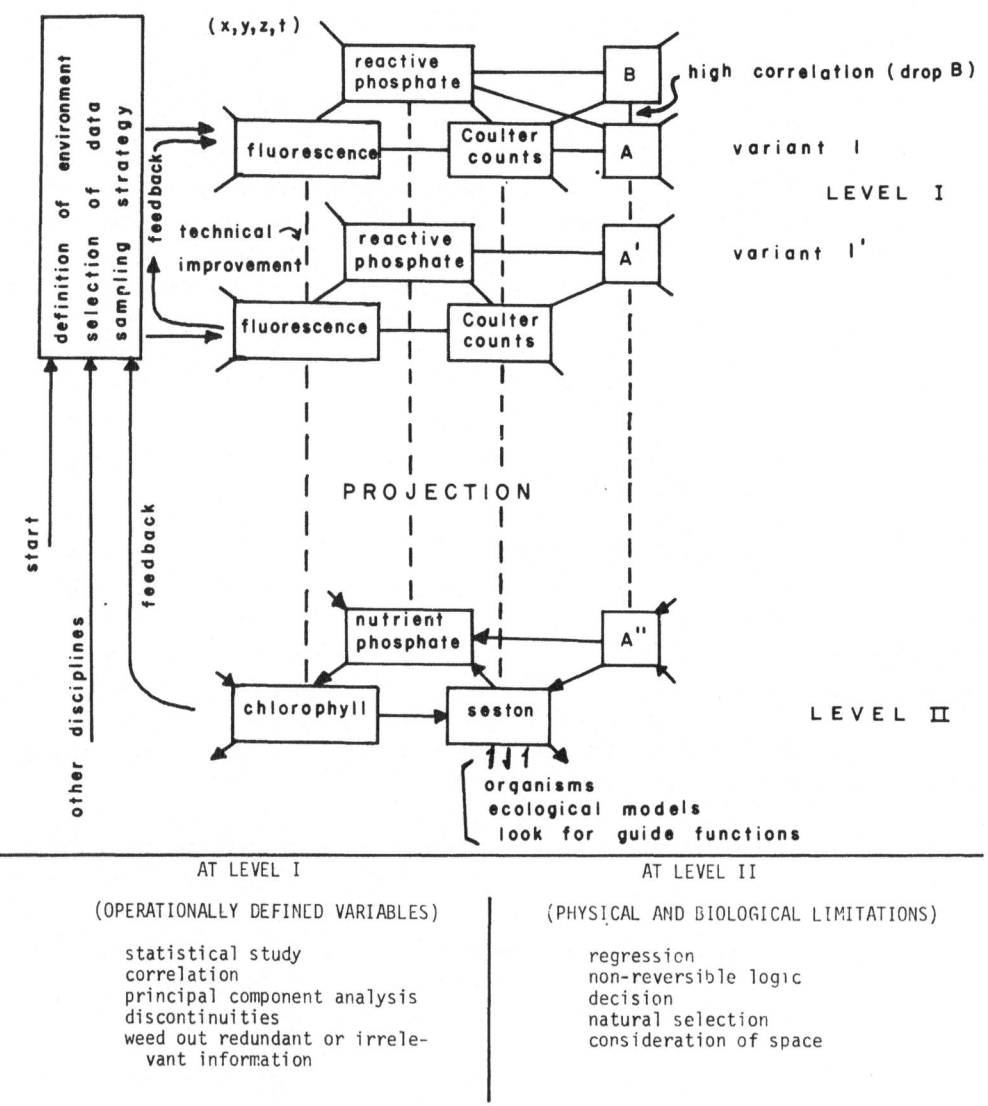

AT LEVEL I	AT LEVEL II
(OPERATIONALLY DEFINED VARIABLES)	(PHYSICAL AND BIOLOGICAL LIMITATIONS)
statistical study	regression
correlation	non-reversible logic
principal component analysis	decision
discontinuities	natural selection
weed out redundant or irrele-	consideration of space
vant information	

FIGURE 68. Levels of conceptualization of environmental data.

This may be fine for physical oceanography, but concerning biological oceanography there is something that bothers me. Biological variables are much more dependent on personal or subjective factors (identification of species, measure of primary production, and so on) and rarely are there large sets of data that could be considered uniform and given the same label or put in the same box. There are reasons for delaying the inclusion of such information in the data banks. But I believe that the following considerations are warranted: With the data corresponding to the usual concepts of biologists, it is possible to construct a similar sort of network of relations between boxes, except that here the links can be represented as arrows, meaning some relation of cause and effect and not simply statistical correlations. Although variables may not be the same in Levels I and II, there is some correspondence between the data that are used in rapid surveys and eventually go in the data bank, and the variables used in the constructions of ecologists; for instance, a correspondence between fluorescence and chlorophyll content, or between Coulter counts and seston. Considering the whole nets of relationships, one of them can be considered as projected upon the other. In other words, there is one level of information that can go to the data bank, and a level of more or less quantifiable information that biologists use in their constructs. Now I think that one level can be mapped on the other; and of course, the biological model (Level II) can feed back information for further improvement of Level I, and this level can also receive influences from parallel models. What I mean is that, in my opinion, data banks should not be enlarged to include all sort of disparate information, but should be sensitive to improvements made in parallel levels of conceptualization that are dependent from data banks for information.

I feel that there is an important methodological difference between both levels (I and II), as suggested as the bottom of Figure 68. In conclusion, I believe that a World Data Bank can only cope at present with data of my Level I, but has to acknowledge the existence of Level II.

PEACHEY—Well, Mr. Chairman, you will be glad to know I'm at a loss for words. Because I am trying to see where information analysis fits in and where you have to draw the line in interpretation if you are ever going to pull together a set of guidelines. It seems to me that perhaps some of the contributors here haven't done justice to the many scientists working in operational settings.

I think we have got to find some way in which the richness of stored routine data, at some of the levels that have been talked about, is available to users to do what they will do, with or without constraints attached to it, and call that the total data system. If you go to the engineering level then it seems to me that you are talking about management, not just data management. I would like to know from Dr. Oppenheimer what he envisages in this package, which would help here.

OPPENHEIMER—I think that Brogden and Lighthart clearly defined the complexity of data collection and use and this part of the Conference is to show that if we are going to have a total system we have to have some understanding of what the total system will be used for. I think that any step forward today on data management is a good step. I don't think that our scientific or management society is ready for the full step. Confucious says a thousand—mile journey starts with a single step and I feel that we are at that point today.

I know that millions of dollars are being spent to obtain new data and believe that we must build a consorted data base which will be useful to a wide variety of the scientific sector. However, we seem to be going farther and farther astray. More data is gathered today than is being placed in a retrievable form by the total scientific community. A step to move data and data systems closer together is the objective of this Conference.

PEACHEY—I think you, Mr. Chairman, made a remark earlier that it was possible to get a long way with the most elementary arrangements for the exchange of data without undue standardization of one kind or another. That, I can understand, where the data remains relatively non-projected and non-predictive. However, it seems to me that when it is already in the form of a model it has become basically a document and would have to be melted down again before you could merge or use it. I cannot understand how to accommodate models in a total systems concept.

OPPENHEIMER—Only as a means to consolidate information. A model is one method to validate real information. A model can be converted from a mathematical to a predictive model by comparing real data or fitting the model to a real situation. Therefore, we must use real data bases with models as a part of a total environmental information triangle.

KOHNKE—I agree with Professor Oppenheimer. I, too, understand the model as an inevitable and absolutely necessary continuation of the field data. If we think of the environment as of a system which is influenced by external forces (tide generating forces, radiation, etc.) — we call this input — energies are transformed in this system. What is measured (tidal wave, chlorophyll, oxygen, etc.) — the output — is the system's reaction to the input. The measuring itself is a mere description of what has been observed. It does not yet give an indication of what virtually happens within the system. This is the task of the model; only the model gives an analysis of the system's processes. It is quite obvious that the modelers cannot do without observed values either, whether they need them as an input for predicting their system's development or because coefficients have to be empirically derived from the observed values; without these coefficients the model would not function.

Therefore, it is very important that the researchers tell the data managers what they expect of a data center for their studies. This has to be taken into consideration at any rate if, some day, the collected data and information are to be utilized reasonably. I think it is relatively easy to define the users' requirements in the marine physical and chemical fields; these data are quantitative values. On the contrary, it is very difficult to achieve agreements on the type of data or information which should be archived in marine geology or in hydrobiology. Also, in marine geology, there is no common view on what type of data should be put in a data system. Some people cannot be persuaded as to the usefulness of such a data system at all. Hence, I would be glad if the research scientists attending this Conference would contribute to a solution to this problem.

SAYDAM—Mr. Chairman, may I take this opportunity to formalize a model to some simple extent. The models and data base, I think, can be analyzed and brought into unified perspective.

First of all what is modeling? Modeling is a set of equations that describe a phenomenon or a set of different phenomena. Usually, you take a box and in each direction you try to write the inflows and outflows and what's going inside the box. For instance, say "d" is the differential operator, "P" is the variable you are investigating, and "Px" and "Py" are the derivatives of "P" in the "x, y" direction, respectively. Therefore d [APx, BPy] must be equal to time rate of change within the system, "Pt." A biologist can write a system like this with "A," "B" and "C" constants; these may be also within the differential operator, so here we have an equation:

$$d \; [APx, \; BPy] \; = \; CPt$$

This is the simpliest form of a mathematical model describing a certain transient phenomena.

But we don't know these coefficients. We can't solve these or any set of values of "A," "B" and "C." This is why we go to modeling. Next we utilize the data base. I will show you what a data base can do for us. We must consider the past, the present and the future (see Figure 69). This is what we must play around with in these models. We have to take the past performance, past behavior of each one of these systems. We know this from the data in the data base because the data base gives us what has happened in the past until today. If we know that behavior, we know the past. We can take the model and match this pattern which we call "history matching." Once we have matched the history, the solid line in Figure 69 is the actual behavior and the dotted line is my model — the predictive computer model. Now when I match, this model becomes a predictive model of this phenomena. From then on I can predict well into the future. This is the main purpose of the model, to be predictive. A model without any predictive capability is nothing at all. It's an elaborate mathematical equation and nothing more than that.

In order to make it meaningful, you must put things into a simulator fashion — a program that simulates the system. How do we go about getting this data? In talking to Dr. Oppenheimer we have been discussing how the data explosion is occurring, where we are, and how we are.   I will draw another graph to illustrate (see Figure 70). Let the horizontal scale be the time and let the vertical be the bits and log scale. The data have been increasing as indicated by the solid line, but our understanding of data, our analyzing capability, our putting this data into meaningful fashion is only increasing slowly, as indicated by the dotted line.

Dr. Oppenheimer assured me that every year $5,000,000,000, wasn't that it Dr. Oppenheimer?

OPPENHEIMER—About $5,000,000,000.

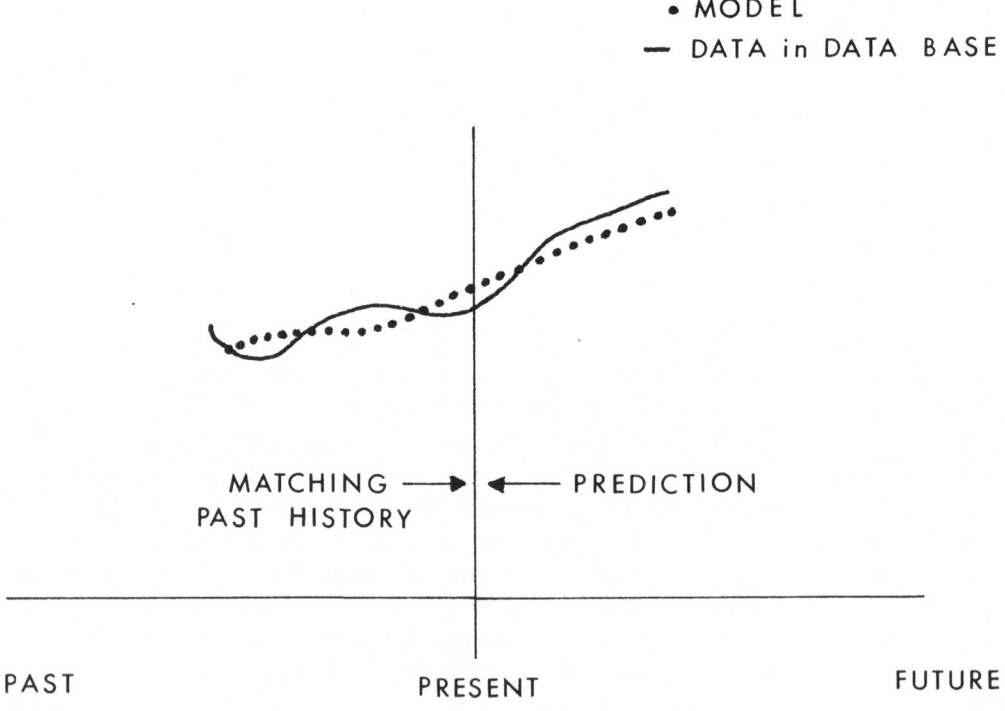

FIGURE 69. Matching history and prediction of future behaviour.

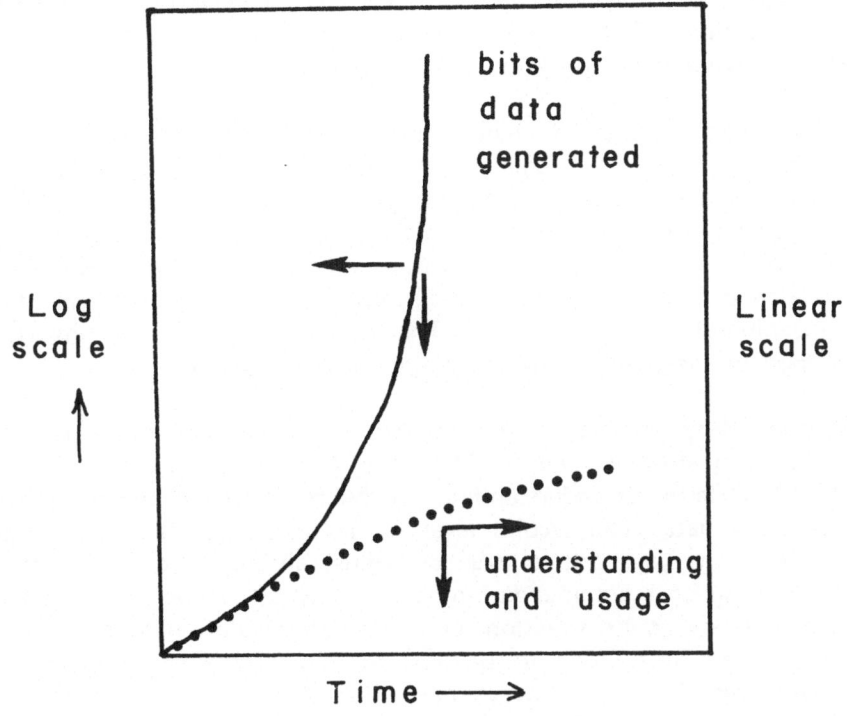

FIGURE 70. Relation between bits of data generated and our understanding and usage.

SAYDAM—About $5,000,000,000 have been spent on collecting this data. But this is a lost cause because we are not able to analyze it. We are simply collecting it. The importance of the data base—the data center—is to close this gap between collecting and analyzing, to shift this lower curve upwards.

I agree entirely with Dr. Kohnke that each discipline should solidify and indicate what types of data it wants. Because each discipline will come up with its own set of requirements and, therefore, the data base must not only reflect one set of data but the different sets of data from different disciplines. But again, research scientists must know which is where, hence the importance of data descriptors and the protocols that describe the data.

I would like to emphasize the following. We have been talking about managing data. But do we really want to manage data per se, or do we want to manage data in order to manage the environment properly? I think the latter is the case. Therefore, when we think in terms of this environmental management, data management becomes simply a subset of it. We must never forget the bigger picture. Another subset, I think, is environmental research utilizing this portion of the data, you can call this "D1." Another research program may utilize another portion, "D2," and so on.

But in order to explain environment management, you have to put all these little circles into Boolean logic which must be connected by AND's, OR's, or NOT's, etc. In order to get meaningful or more relevant results out of this discussion, I again suggest that we look at the broader picture, not only at data management itself. I see data management not as the final objective, (this is the final objective for some people, yes.) but as an initial objective—that of providing meaningful data to models. The final objective is to provide this meaningful data to the people who make qualitative decisions.

Not everybody goes to models to describe nature, but some people like to analyze in a a qualitative manner. Therefore, I can distinguish three levels of activity. The first level is understanding the behavior within the environment. For this we need data. The second level is the predictive level, the modeling level. Modeling is always a critical task and must always be critical. The third level is the advising or recommendation level. After all, if we do not take any political stand, or any advisory action, or any recommendation to the respective governments, we are not going to be able to control the environment at all, because environmental control always has political implications behind it, and thus requires political action for implementation. To trigger this action we have to make recommendations. For these three different sets of activities, we collect and manage data.

KOHNKE—Thank you, Professor Saydam. I agree with you and I think that was a very good explanation of the connection we have to make between the theory and the field measurement. I have the feeling that most of our colleagues will agree with you. I would like to ask you to develop some ideas of how we can approach this total system in practice. Which steps have to be taken to meet the requirements of scientists to come to a better understanding of their specific measurements and their specific results? That is what we are here for today.

What you show in Figure 69 is exactly what I call the black box. What we are measuring is, in any case, the output, and perhaps we know something about the input. Let's take the example I already mentioned, the upwelling. We know the upwelling, we know the input—the tradewinds—but what really happens physically can only be explained by a model. The model is elucidating the black box.

I would like to ask some of our colleagues here to develop some thoughts towards practical steps for the development of the total system. Would anyone be willing to contribute to this point?

GORE—May I make a contribution before you ask for that?

KOHNKE—Yes.

GORE—I have enjoyed the hospitality of our host and hostess, and indeed of this country in the last few days, and it's with reluctance that I must question the need for a total data base. I find myself increasingly in a position of wondering whether this is a valid objective. I think the crucial thing that matters in scientific affairs and indeed in the management of man's affairs is ideas—the synthesis of the impressions that we receive from the world around us, and the interpretation of these in a manner which our fellow man can understand and appreciate.

I do not think it is necessary for us to preserve indefinitely every single item of those impressions that we have caught. John Peachey gave light to this when he said that he may be phoned up and told, "Don't give me the data, give me the picture." It is indeed the picture that we need, not the data. This is not to say that individual scientists or groups of scientists don't need data bases. But as far as a total data base is concerned, I very seriously doubt its necessity. My professor used to say when a piece of work was done, "What did he get out of it?" We all knew exactly what that question meant—did he get anything out of it? Of course he got enormous numbers of data, but that wasn't the issue. What he was talking about was what we got out of it as far as contributing to the understanding of our environment. Those are the things that need to be on record. If we need to check the ideas, (those ideas which have been put forward, what our colleagues and forebears did get out of it) then we can go and make other measurements, and not use their measurements at all. I submit that we should seriously consider this proposition. Forgive me for any apparent inhospitability.

KOHNKE—May I ask how you will get the picture without having data?

PEACHEY—You get it because people know that there's something wrong. The concern about what we might call research data is to my mind disproportionate to what is needed operationally to meet our obligations to keep nasty things away from people and to retain pleasant things. Ordinary people can now read and are saying I expect scientific observation to lead to proposed, if not actual, remedy.

What worries me insomuch about this talk on environmental management is that it's basically related to an assumed obligation that we do have to know everything about our universe. That was part of the space era philosophy, there was a total commitment to know everything, to push forward. But now we are coming down from the mountain and finding the world in which we live now as our first priority. And for that perhaps a comparatively low level of information is needed. You, as scientists, have got to re-earn your credibility in modern society.

Wouldn't it be easier to build this system's thinking about real needs of real people in real situations and then tackle the very difficult global phenomena in time, taking the precaution of banking relevant data in an orderly fashion in the meanwhile?

OPPENHEIMER—Well, I thought that's what we were talking about.

PEACHEY—But you have given emphasis to a very complex number of long term interests.

OPPENHEIMER—The point that Saydam made, that the past is the key to the future, is a concept that the geologists have used for years. The ecologists are just now beginning to get enough information in their files to do the same thing. The concept is that there is no such thing as present, because it's transient and our environment is complex. We cannot explain the environment in simple terms.

The crises that we have with the environment can be very easily overcome with a good data base. We can clean up our rivers and our streams and so on very quickly if we want to, we can stop throwing beer cans along the side of the road, as that's a very simple personal task, but the problem is really one that Dr. Saydam brought out and we keep bringing up, and that is that there is a dollar sign or an economy sign to environmental control and data systems. We are being forced today to do things that, because of insufficient knowledge, are costing us needless money. Therefore we must go to a coordinated data base and a data management system.

Can we place environmental data in an orderly fashion for retrieval? We are discussing the question of interpretation as to why we need an orderly data system. I don't think that Dr. Gore and I are contrary in philosophy. I am sure he has the same need for a data base as I do as an ecologist. The question is how far can we go to provide coordinative data systems.

BROGDEN—To get down to specifics, I would like to start everyone thinking about how a total system might evolve. I think the first thing you would pick up would be your simple environmentally-critical, justifiable variables that you have been able to agree on; for instance the 24 variables that will be measured. During the evolution of the system I think this is the first thing you pick up. The second thing you would pick up would be those variables which have been demonstrated to be already internationally agreeable such as the meteorological network and the oceanographic network—those parameters which have already been demonstrated to be shareable and well defined. Then I think we would pick up the variables on land which are similar to the meteorological and oceanographic parameters, and which are a little bit harder to describe. Next I think we would pick up the biological and geological descriptions of the environment which are the data related to the existence of objects within the environment, for instance the pattern of plants or pattern of soil type occurrences. And the next step, which would be much harder, would be to begin to pick up the processes. From there I'm not sure how the evolution would occur, but I think it would occur along these lines.

CUTBILL—I'm going to stir the anthills. There is obviously a difference of opinion about the desirability of the whole thing. I want to pick up something that Dr. Helms said that seems to have been jumped over. I will rephrase it; there is a point of view which says that a well organized body of environmental data constitutes a hazard to the environment. Of course we have got a vested interest in saying that is nonsense, as we are talking about systems which are either for our personal use, if we are scientists, or are for the administrator's use. And we are looking toward some sort of recommendations again. Most of the discussion is in terms of systems, practical ways of getting there. However, somewhere along the line I believe we must make some pretty positive statement about why this is all a good thing. I just want to get this anti-view onto the record so that this meeting can make up its mind positively and absolutely and reject it if it wants to, with some reasons.

RANNESTAD—I think it is inevitable that we will have a data bank system; it will evolve regardless of whether you like it or not. Thus, our main problem is to define certain descriptors which will enable the scientists to use the data in the system.

PEACHEY—Yes, but you have made the major concession that was made in getting GEMS through; namely, that the system was to be in effect a voluntary network with full regard to national interests and existing capabilities.

This concession was made in order to get over the problems mentioned earlier in discussing Dr. Helm's paper.

In addition to the loose kind of general monitoring network envisaged, it is hoped that part of the system could, by agreement, be more rigorously developed so as to insure the adequate monitoring of a few pollutants or other environmental indicators which are distributed in a truly global fashion and which would be useful in determining man's effect on the biosphere as a whole. This word "system" in the context of GEMS now means a "network" and we perhaps ought to use that term. That then brings in this question of the whole descriptor business and the formats that sustain the ensuing network arrangements.

RANNESTAD—I look at the concept of a total environmental data system somewhat differently. A total environmental data system may be for the Corpus Christi Bay, for the State of Texas, for the whole U.S., or for the whole world. Regardless of how big the system is, if it is going to be useful it must have a common denominator, it must have a standardized data form so that you know exactly what has been measured and how it was measured. Thus, the essence of this meeting is to agree on definitions—systems, components, descriptors, etc.

It has been stated that only part of the necessary data is available for the whole world. This is true, and it can, therefore, only be used for limited investigations. In local areas, however, you have more complete data and by incorporating all this data in your "local environmental system data bank" you are able to perform more complex investigations. I think the essence is that we do not have to consider a system covering the whole world when agreeing on definitions, but we have to be sure that the definitions will enable us to interconnect different systems. In other words, insure that the definitions agreed upon will make systems interoperable.

OPPENHEIMER—I think that Dr. Rannestad has pointed out something that evolved during the development of the Conference, and that is that we never conceived the total system as being one great computer. The term "system" was taken as a concept, perhaps the best network, or system descriptors which would allow us to describe the environment.

I should like again to point out that whether we like it or not there are data being accumulated that are extremely valuable in terms of cost as well as importance. These data are, to a great extent, except for those systems that Brogden mentioned such as the oceanographic system and so on, in somewhat of a state of chaos today. If data systems are not in a state of chaos, they are in very small, non-compatible areas, and it takes money and effort to consolidate the data.

The one question that has been emphasized in this Conference is, "Can we place our data in a more easily retrievable form—data that are already being collected, entered into small systems, into drawers, on pieces of paper, onto punch cards, onto IBM cards, into tape libraries, into international or national data centers?" The world has a right to that data. There are millions of bits of data accumulated that are not only valauble to the individual collector, but also to the person who will use such a base to understand the environment and assist in the decisions that have got to be made regarding the use of our environment.

I hope I did not give the impression that I am trying to provide a monolithic data structure where somebody walks into a big building and there around him are all the data in the world. I think this is absolutely inconceivable. That was not the original idea of a total system. A total system is a system in which we can place the data for retrieval. I think that we mentioned the network system as probably being the logical conclusion.

KOHNKE—I agree, because I think that a data center too big could paralyze itself. I mentioned this the first day, therefore, we should have a certain de-centralization. But we have to define the links between these different data banks and their data systems. The most reasonable way to come to a better utilization of data throughout the world is to improve the interchangeability of that data. We don't have to recommend how to collect more data, only how to make the existing value available to the scientific community. I have the feeling that there is total agreement of the participants of this Conference that the environmental data and information system is not identical with a rather big shop shifting data from one side to another.

LOUDON—It seems to me that our main problem is that some of our terms have not been very clearly defined, and perhaps to some extent we are getting carried away by our own propaganda. Some of the arguments do not seem to be completely logical to me. The diagram by Dr. Saydam (Figure 70) demonstrated that increasingly more data are being collected, and are increasingly outstripping the scientists' capacity to assimilate them. Improving data communication and data linkages would seem to me to aggravate this problem. Systems concepts would indicate that balance could be restored by reducing data collection or improving the abilities to interpret the data.

The other thing that strikes me as being a bit odd, in terms of system thinking, is the title of the session today, "Towards a Total Information System." I had supposed that there is, always has been, and always will be, a total information system comprising all the elements, most quite conventional and well established, of obtaining, analyzing, communicating and using information. It may not be a perfect system, and presumably never will be, but it is always there and always with us. The "total system" to me means all of these activities and their relationships considered together, and is not something one moves toward. Dr. Citron, who unfortunately is not here at the moment, made me feel rather uneasy this afternoon, when he stressed the vital significance of our deliberations, leaving me with the impression that he felt that the future of the world was in the hands of those around this table. From which of us would _you_ buy a used car?

Surely, all we can hope to do with new techniques and computer methods is to make marginal improvements to the present system. If part of the problem is that there is too much data to handle, then I suggest that we concentrate not on adding to the surplus, but on seeking ways in which we can more effectively use existing data.

KOHNKE—I am sorry, but I cannot follow you when you say that we already have a total system. What we have are so many data distributed over so many countries, so many institutes and labs. What we need is a suitable organization to link the existing data. Today the arrangements are absolutely insufficient.

LOUDON—The problem is the definition of the word "system." I entirely agree with what you are suggesting. I disagree with the word, but I agree with your idea.

SAYDAM—It seems to me that the environment today is a system which requires immediate action. Tomorrow may be too late, or ten days from now may be too late. Therefore, immediate assessment of data is of paramount importance, you cannot lag back five or ten years. Analyzing ten year old data will not help you solve today's problem. When I drew that chart and showed the lost data I meant that you have to analyze data immediately otherwise it won't help.

Let me give you an example, if you cannot form a proper data base initially, it will be very costly for you to do it later. If I am not mistaken, it has been estimated that the Library of Congress of the United States once tried to estimate the cost of putting their data into a computer. For simply codifying the current data, the estimated cost was on the order of $180,000,000, and they gave up the project. They can give up the project because their need may not be immediate, or they can rectify it tomorrow, but it seems to me that you cannot wait as far as the environment is concerned.

OPPENHEIMER—I think we can emphasize that too.

SAYDAM—I have one more point to make. I do not agree with Dr. Loudon in saying that we have a system now and we can make marginal improvements to it. We can't. On the one hand we have a mechanical system which is lagging behind, on the other hand, we want to increase the computerized data base system. These are two different operations as you cannot improve the current system marginally to arrive at the better data base system.

LOUDON—The introduction of computer methods is a marginal improvement on the system.

PEACHEY—Mr. Chairman, when I was involved in information science research and development, we were constantly being told that we should think in terms of total systems so I guess that this was an information science or management term which basically implied a mixed multi-phased approach. I doubt if it had any primary institutional implications. But what I want to know, and perhaps Dr. Berg can answer my question, is did the total systems concept really take root in the field of scientific and technical information?

BERG—I'm not sure I can answer your question but I will try. Let me make one or two comments first. I think there is a confusion on the definition of words, specifically "systems," but I think that is not the only problem. I think a question to be asked is, "are we really trying to put one data base together or not, or one system together or not?" I think many people feel we should not try to put together one total system, or one total data base, and I would subscribe to that. Data for such systems are too disparate.

We have not talked about any of the other types of data that I think are important, such as economic, social, and all the other fields related to managing the environment, which should be part of the data base as well as physical measurements. We cannot talk about effects on the environment without talking about social, economic, and aesthetic impacts. When you include those the number and character of the data items are increased so much that it is not possible to put it all in one data base. The data are quite different. Some are numbers and some are not; these latter may be used to describe part of the data.

Additionally, there are qualitative assessments that could be involved. For instance, having beer cans along the road may produce very little harm physically, in terms of destruction to the environment, but aesthetically it's quite bad. How do you describe that and how do you characterize that in the system itself? I don't think you can completely characterize such concepts. I think that's a problem that you cannot solve to the extent that more quantitative problems can be dealt with.

I think there is a problem in saying that we have a system now (which I think is true if you want to use the term for very broad purposes), but it's not what we are trying to get to. We are trying here to get to an organized set of procedures that will enable us to interchange data in a better way than we do now. I don't know if people would agree with that or not, but I think it's important that we have some foundation, some agreement about terminology before we get too far off on a side issue; we seem to be veering off to side issues somewhat. If we don't agree on the terms, then I think we are in trouble because some of us may be talking about one aspect and another person is misinterpreting the comments and talking about another issue.

At any rate, we should come down to a discussion of types of improved procedures for the interchange of data that are needed. I think we also have to examine the scope of this problem. Are we talking simply about measured data, physical data, or are we talking about other data as well? If we also consider data management, then we may be involved with several aspects of management—systems in addition to data management. At any rate, if we get into total systems discussions, I think we are in trouble. I think if we try to take in too many concepts we have a problem. Is there any agreement about that or is there any obvious disagreement? I think we should answer questions right at this point, or tomorrow we'll have no recommendations.

RANNESTAD—It has been mentioned that areas such as libraries have agreed on a system. The herbariums are today, as a result of the Kew meeting, working on a set of descriptors which will be agreed as a minimum standard. What we are faced with here is considerably more difficult. When the areas mentioned include scientists working in the same field, it should not be too difficult to agree on a certain set of descriptors. But here we have a group of scientists and administrators working in different fields and it, therefore, makes it much more difficult to come to an agreement on details. Let us not consider at the present time a set of descriptors, but rather a system concept. Let us, furthermore, forget our specific projects, our specific way of working, and try to look at the problem as a team of different people covering different fields with the aim of defining a total information system that can be used by all of us.

I agree with the statement that we cannot collect all the data in the world and put it into one room. This would be ridiculous. My concept of a total system is, as mentioned before, different and more flexible. A total environmental system (sub-system) may, in my view, cover a very limited field or a limited area, but include all necessary information to do your job. The necessary data may not be in one data bank, but you must be able to retrieve the necessary data from other data banks included in the system. Several such sub-systems may be interconnected to form a larger system for more complex investigations.

BERG—I think you will need different types of descriptors though.

RANNESTAD—Yes, you most likely will. But the different sets of descriptors must be based on a common concept, so that you can manage to connect the data banks together.

CUTBILL—I would like to try to address this subject. We are now in the practical area where we can make a good recommendation.

Although some of the things we measure are local considerations, a lot of the things about what we measure are not local. Because there are a lot of quite broad concepts about representation of accuracy, etc., that can be described in general terms, we should be able to standardize particular parameters or the general terms for a range of sets of data. Any attempt to agree on a convention for describing background information which will cover a wide range of data has immediately improved its communicability. It's just a recognition of what is going on, of the fact that seemingly totally different things around us have common qualities, and if we describe these in the same way we're in business. I'll submit this to a drafting committee, because I feel here is something on which we should come up with a recommendation.

OPPENHEIMER—I hate to have our meeting break down for want of a definition of a word. I think we are putting too much emphasis on "system." According to the dictionary, a system is a set or arrangement of things so related or connected as to form a unit. And I think it was in this sense that we originally used the word. Perhaps we should have used "unity" rather than "system." Maybe some people thought we were trying to generate another solar system; however, a library is a system, our telephone services are a system, and I think that we need a system for our data bases.

It's very true that we cannot get a set of descriptors which everybody agrees to, but there should be certain common denominators which can be adopted. These would then form a basis of a system. We can also describe the mechanical things that we want as part of the system. The data program must coordinate major areas so that we can have an interchange within a global data system. With the state of data compressibility and software compatability, it is possible to provide unity. If each small, existing system is continued with its own software, we will go farther and farther away from the ability to have on-line capabilities with different data systems.

We need to develop a concept of a system somewhat as in the dictionary; a unitizing of data so that it can be useful to a wider sector of the community at a smaller and smaller cost. A library is an excellent system, but it is only as good as its system for retrieval. A chemical abstract is a simplified version of a data base system. But it is one step down the line, as it gives us the freedom to scan without going back to the shelf. However, we need to carry the system to the point where we not only have a summary, but also have the discrete data points as well. Our total information system will be made up of data sources.

ROSENFELD—Systems provide for the exchange of data but I think again, we are at two levels. You can't exchange very dissimilar data.

OPPENHEIMER—Right now we can merge many data bases, but it takes extra software steps. What I should like to emphasize is that if we design a system now, we may avoid the multiplication of non-compatible data systems in the fugure.

RANNESTAD—This is exactly the point; we must define a concept which facilitates the exchange of data without the necessity of translation programs.

OPPENHEIMER—There are hundreds of scientists, each with individual ideas and programs that will continue setting up their own little system. What we need are guidelines for systems that would be more easily compatible and accessible. A scientist may not have the time, in some cases, to spend days or weeks looking for data in the literature and in other data bases, so it is easier to ignore past data. However, when you ignore data you run the risk of misrepresentation; that is dangerous.

I think Dr. Kohnke, it would be appropriate to summarize this session as we have had a very fruitful afternoon.

KOHNKE—Yes, I agree. I will try to cover the main points. There was hardly any disparity of views about the necessity of making the existing data and information available to the interested general public. No unanimous agreement could, however, be reached with regard to the priority that should be given to the establishment of an organized environmental data management system.

Two groups of interests could be distinguished here. One of them consisted of the administrators whose responsibility, among other things, is to protect the environment against damage, if necessary, by legislative means. This group has to make its decisions not only according to scientific necessities; but also by taking into consideration factors of social and economic policy. This group is interested in more general information and final scientific results. The second group consists of pure scientists whose objective, in the first place, is the original data, in order to recognize the natural laws in environmental processes. This, however, is part of the basic research and is initially not carried out with a view to practical utilization.

Both groups evidently expect different things of an environmental management system. For the recognition of exact natural laws the greatest possible amount of data and information must be available to the scientist. Without the utilization of scientific knowledge, however, every action taken to protect the environment is bound to be unsatisfactory. No national or international monitoring programs should blind us to this fact. Environmental monitoring per se is not yet an answer. I think it is even dangerous to propagate such programs publicly. This might easily give the public the impression that, together with the programs, the environmental protection has started too. Environmental protection, however, can only begin after a thorough analysis of the data collected in the monitoring programs. In order to be in a position to implement this as effectively as possible, the basis must be created for complex interdisciplinary investigations whose results would be of assistance in deciding which influences to the environment are still permissible and which have to be absolutely avoided.

In this sense, the effectiveness of a data management system was generally admitted. It further came out that it is absolutely inconceivable that the data system to be aspired to will have a monolithic structure like a single warehouse where the scientists from all over the world can receive data and information.

What is much more promising is to interconnect individual data systems, data centers or data exchange networks. This interconnection does not necessarily have to be done by the "on-line method," although some participants in the discussion thought this would be useful. Whether such an on-line operation is to be established should depend in each instance on whether the data have to be available in real time or in non-real time.

Regardless of whether computer systems are interconnected directly or whether the data re-exchange on suitable media, the application of certain standards has to be aimed at. Both the computer systems and the contents of the data systems, i.e., the parameters themselves, have to be compatible. For the computers it can be assumed that their compatibility will improve in the future. As to the comparability of the observed values, we are, in certain fields, still far from a standardization of recording and analysis methods. The question of which parameters are relevant for the recognition of natural laws and the definition of tolerance limits for the impact on the environment will be asked over and over again. The necessity of adaptability to changing situations is another reason the "monolithic concept" is out of the question.

The reliability of the data themselves must, therefore, become obvious to the secondary user from supplementary information. Otherwise, it may be feared that the data available do not receive the trustworthiness which is necessary for their utilization.

It seems unreasonable to propose the utilization of a standardized format for data processing. On the contrary, each lab or data center should develop its own formats, taking into consideration the standard criteria that are generally regarded as necessary for an exchange. Then it would be possible to carry out the non-real-time exchange of data in a flexible, standardized exchange format. If on both the national and international level an agreement could be reached on the exchange standards, this would mean a big step towards an environmental data management system.

IX. CHAIRMAN'S COMMENTS AND RECOMMENDATIONS

I have taken the liberty to summarize various concepts discussed during the Conference, particularly the last morning of the Conference which is not included in the text. The last morning was a free and open discussion devoted to summarizing, developing the basic concept of an Environmental Data System (EDS), outlining the basic requirements of an EDS and developing a list of recommendations for an EDS which were to be sent to NATO and the United Nations for their information.

A summary of the recommended global EDS is presented in the following outline:

1.   Purpose:

To provide the capability for the improvement of the global exchange of environmental information and the preservation of environmental data.

2.   Basic Premise:

A.   There is a need for an environmental Data System

B.   Computer technology presently available has the capacity to implement an EDS.

C.   Several aspects of regional environmental data are already successfully compiled and exchanged on an international basis.

3.   Recommendations:

A.   Interfacing between the proposed system and various users requires a system of compatible environmental descriptors and documentation including:

1)   an environmental dictionary or thesaurus of terms in use, and

2)   a set of minimum descriptors for all data.

222

B.   The system must provide the interface and network capability to provide scheduled access to local, national and international data sources.

C.   The system must respect the integrity of the original data and provide information on the origin, state of verification and availability.

D.   Contribution of data to the system should be analogous to publication in a scientific journal in terms of creating scientific credit.

E.   The cost of any system must be related to the expected economic and social benefits to men and women.

3.   Requirements:

A.   Increase in storage capacity and retrieval speed.

B.   Universal descriptors or language developed with the basic communication characteristics of:

1)   language unity—thesaurus, library and bibliography—general, specific or field oriented

2)   data preservation

3)   a data integrity system—status of verification, availability, etc.

4)   on-line status—plot, cluster, graph, statistical, model

5)   form of credit to data donor

4.   Access Methodology:

A.   Entry features

1)   location of environmental data

a)   geographic—latitude, longitude, grid, etc.

b)   reference to origin—person, project, agency, country, etc.

c)   items measured or described—name of parameter, method of measurement, accuracy and precision, summary of other similar pertinent data files

2)   file codes

a)   bank location

b)   contact person or code

c)   computer and language used

d)   transfer method

e)   integrating requirements

3)   other information

a)   publication restrictions

b)   method of referencing data

c)   cost of acquisition

The discussions throughout the Conference were varied and reflected the thoughts of the participants as they listened and provided input from their experience and knowledge. Through these discussions we learned that there are about as many types of data management philosophies as there were participants and each participant had a data file or program to prove his point.

One half day of the Conference was devoted to computer demonstrations where there was an opportunity to inspect and work with several different data systems. Later discussions centered on computer advances and the resulting possibilities for sharing data files in the future. Thus, a very wide scope of philosophy and detail of existing computer systems was presented.

As Conference Chairman, it was apparent to me that one of the primary problems our group faced relating to environmental data was a communication problem. During the evolution of data information systems, various natural and computer languages have evolved. In the discussions, these language differences caused considerable difficulty because of the different identities or meanings given individual descriptors or environmental characterizations. However, as seen from the point of view of the Chairman, there seemed to be a commonality between conversations that could resolve these language difficulties by building a common environmental language or thesaurus.

The underlying theme of this common baseline between participants was this: I have a computer system that works well for the purpose for which it was developed. I see that you have also developed a similar system but with a different purpose. Our programs are different, however, the mechanisms are the same, and we have used a common denominator—the computer—to reach our individual goals.

It is my opinion as editor, that such an undercurrent merely emphasized that most of the participants had their own computer language and had not really considered the commonality that may be possible between systems.

Another point became apparent regarding support for programs. As many of the participants were project leaders, they were all concerned with the cost of a common EDS—both its development and operation. Generally, computer programs are under-supported in comparison with the funds available for the acquisition of new environmental information. IT IS MUCH EASIER TO OBTAIN FUNDS TO CONDUCT BASIC RESEARCH INVOLVING NEW ENVIRONMENTAL INFORMATION THAN IT IS TO PROCESS THE INFORMATION AND MERGE IT INTO SOME COMMON SYSTEM. Therefore, the term "cost" was introduced early into the discussion. The concept was basic and can be emphasized simply, "if we do not have sufficient funds to process existing information, how can be obtain the additional funds needed to merge our data systems?" A very valid question that may have produced considerable negativeness throughout the Conference.

Dr. Helms presented a most magnificent talk related to computer engineering capabilities. Most of the participants were concerned with the magnitude of the total environmental data base as related to an EDS. The question was asked, "How can we mathematically or mechanically contain a world data system?" Dr. Helms, quite appropriately, discussed the issue from an engineering point of view to show that new developments underway could accept the total data base of a general system. He indicated that the technology exists to place the data from a 100,000 tape reel storage area into a box $1.2 \times 1.7 \times 1.7$ meters (referred to as the "little black box"), with a compressibility that is equivalent to replace a 2,400 ft. tape with one square inch of tape surface. He emphasized that the technology was available, it merely needed implementation.

As we envision a total environmental system involving a large mass of environmental information, the little black box concept that Helms mentioned came as a breath of fresh air. If the capabilities for such data compression are real, then we have no hardware problems for world data assimilation. We merely have to conceive of a computer communication system that is compatible with the large data base. However, Helms also emphasized that the primary problem in moving to a network or EDS would be related to program management. It would be difficult to interface existing programming action and requirements with the hardware needed for a global network system.

If the little black box concept is implemented, it would provide the required compressibility of an intermediate storage system compatible with existing computer programs and could allow an infinite capacity for interfacing between data systems. While I do not wish to dwell on the little black box, it is important to show that the computer may be adaptable to the network concept that was discussed during the Conference.

A question that was repeatedly raised throughout the Conference concerned the basic requirements of a total environmental data system. We heard about the weather data system and its capabilities, oceanic data systems from both sides of the Atlantic, specialized environmental systems designed to show man's impact on the environment and to characterize localized biomes, and bibliographic systems such as Chemical Abstracts. Theoretical systems that suggested the essence of a total data system were outlined.

The term "environmental system" was discussed as it is an important concept of any program that is to be developed for the consolidation of environmental data. For purposes discussed here the word "system" is all-encompassing. It should not be confused with a telecommunication, a world weather service, or others such as computer hardware systems of any single manufacturer containing central and peripheral units. Thus, the word "system" encompasses the multitude of requirements including the above and in addition, other qualifications such as a unified language of descriptors. The ultimate system would consist of a network of compatible data systems, each of which could be used both for the purpose for which it was originally developed and could interchange data through the various techniques.

It is true that some international systems are currently in use. However, these are generally of some specific character such as oceanographic centers in the USA and Germany. At the same time, numerous coastal data computer programs exist that are not included or interchangeable with the Oceanographic Data Centers. To this must be added the several existing data banks that include shoreline information and upland information which, in a Global environmental sense, are all interrelated.

Here we must proceed carefully. The general concept originating out of the Conference was that a general environmental system must consist of an interlocking network of individual data systems. (And I hasten to add that never was a huge building that contained all the world's data even considered.) Thus, the environmental data system must be capable of accepting a variety of different systems.

Each of the participants gave examples of, or references to, their individual computer programs. One of the important aspects of our Conference was to indicate that such computer systems must have some common denominator and, therefore, must have some common interchange between systems. It is this author's concept, in view of the great need for environmental evaluation, that data systems must be made compatible. The electronic and mechanical capabilities of computer systems are available to allow the interchange and the flexibility for a total world environmental data system or interlocking network.

And, finally, the concept of preservation of environmental data, while not brought out in a special part of the program was brought into the discussion many times during the sessions. It is obvious, because of a lack of storage capacity and characteristics of some continuous recording of data, that much valuable information and data which have cost society for collection have been discarded or lost in some forgotten files.

Therefore, I should like to close my brief summary by quoting Austin's Law, which simply states "Whatever data you purge today is what somebody's going to need tomorrow."

The participants prepared and endorsed unanimously a recommendation to be published and sent to NATO and the United Nations for their information. These recommendations follow:

Final Recommendations Endorsed By Conference Participants

There is a growing concern among the world's environmental scientists, that vast amounts of environmental data are accumulating that are not generally available because of a lack of generalized system for data management. The literature is accumulating more rapidly than the scientist can respond and abstract services do not have sufficient coverage to include complete reference to interdisciplinary environmental information or sources of such information.

It is recognized that most of today's environmental data arises from public funding and therefore data gathering represents a large investment and public trust. Unfortunately, due to the increase in publication activities and cost, most scientific journals restrict the author so that only reduced or summarized information can be made available to the scientist. However, such data banks are self-contained and the systems are not compatible without some serious software adaptation.

During the U.N. Conference on the Human Environment held in Stockholm in June, 1972, while information needs were not discretely provided as a resolution, many resolutions passed mentioned the need for improving the arrangements for pooling and exchanging environmental data. It was to this point that the Science Committee Eco-Science Panel of the North Atlantic Treaty Organization funded a workshop to discuss the issue. This workshop was convened in Houston April 8-11, 1974, where 35 experts in ecology, environmental data management, computer science and information handling, discussed local and international aspects of environmental data management. Because of the total breadth of today's environmental concern we attempted to use the coastal environment as a focus of attention, as processes occurring in this area can relate in principle to most environments.

The experts attending the Conference from 13 countries met in free but organized round table discussions of the state of the art, data bases, computer equipment capabilities and the relation of existing environmental data systems. Considerable time was spent discussing potential unifying concepts and the coordination and linking of present and developing interdisciplinary data systems relating to the environment. Several countries represented, were in the initial stages of developing national data systems and we discussed the opportunity to use existing systems as a basis for new development. To this point five different data systems were demonstrated during the meeting by remote terminal. These data systems represented federal, university, local, state government and industrial systems, all designed to answer questions relative to environmental data.

It was the general conclusion that there exist data systems that could be woven into a global network and that the hardware was available or in the process of development that would accept the scope of global information today and in the future. Such data storage capability information appeased the concern of many participants that we would soon have more information and data than could be contained by any information retrieval system. One half day was spent at the Houston NASA Manned Space Center where the ERTS program was described and the participants were able to see the operations center of the Manned Space Vehicle.

Two of the conferees recently attended the United Nations Session on Global Monitoring and International Referral System for Sources of Environmental Information. Therefore the conferees had the opportunity to learn first-hand the important discussions that took place at those meetings in Nairobe in February and March of 1974.

In the final morning session of the Conference we discussed the aspects of a total system and as a result unanimously endorsed the following statement which was forwarded to the United Nations. The proceedings of the Conference will be published.

"The Participants in this Conference, being experts in ecology, environmental management, environmental data assessment, computer science and information handling; attending in their personal capacities;

Taking into consideration the urgent need of men and women for a decent, secure and peaceful environment;

Acknowledging the support given by the NATO Science Committee which made this Conference possible;

Noting the substantial progress made by the United Nations Environment Program and by similar regional, national, and sectoral equivalents;

Taking due account of substantial advances in data handling methods;

RECOMMEND to the appropriate governmental, intergovernmental, industrial, academic and professional bodies the following observations and suggestions:

1.   The need to provide a systematic set of procedures for organizing environmental data bases and for facilitating access to, or the exchange of data by, the adoption of appropriate communications formats either for environmental data files or for individual environmental data records.

2.   The importance of the emergence within the United Nations Environment Program of the International Referral System for sources of environmental information (IRS), the Global Environmental Monitoring Systems (GEMS), the International Register of Potentially Toxic Chemicals and general ecological inventories (and other similar environmental data collecting, handling and processing activities) to meet the total data and information needs of environmental management programs at the global, regional, national, local, sectoral and scientific levels. This is seen as a first step in the evolutionary approach towards the total systems concept which motivated the organization and successful conclusion of the present Conference.

3.   The importance of the iterative approach in using advanced techniques of data assessment.

4.   The need to promote a proper description of subject matter, methodology, data structure and authenticity on the one hand and an adequate description of location in space and time on the other.

5.   The usefulness of advanced techniques of computer management including on-line access in helping scientists, managers and other data users to establish an effective dialog with appropriate data systems and to help ensure that meaningful questions are developed in the course of interrogating these collections of data. This is the initial priority in the use of on-line facilities for perfecting this dialog notwithstanding future requirements necessary to ensure timeliness and quick response through full on-line facilities.

6.   The need to acknowledge   the dangers inherent in the construction of data-rich files and to reassure individuals, institutions and nations that the threat to the integrity of data or to the sovereignty of national and other concerns can be protected by full respect and adequate arrangements for confidentiality. The evolution of voluntary compatible data networks (a major concern of the Participants in the present Conference) is seen as the most adequate safeguard against these hazards without impairing the best use of accumulated knowledge.

Those to whom these recommendations are addressed are urged to consider seriously the obligations of the data collecting bodies within their area of responsibility and to encourage these bodies to use IRS and GEMS and related data handling initiatives as the basis for developing a general systems design capability so as to push forward to a level of data sharing which will enable a genuine interdisciplinary and transsectoral analysis of environmental pollutants and their effects, of baseline data and of natural resources. This will assist materially and socially in the protection and development of the human environment.

In these suggestions and observations, the Participants in the Conference recognize that significant advances towards the provision of total data systems to assist in the assessment and management of the environment have already been achieved in relation to certain ecosystems and sectoral areas of interest. Nevertheless, the Participants recognize that this process has only just begun and it is their duty to draw attention to the very rich rewards, socially, environmentally and economically which will follow if this attitude is adopted in all fields of environmental management. Houston, Texas."

April 11, 1974

GLOSSARY OF TERMS

ACI   — Applications Consultants, Inc., Houston, Tx.

ADP   — Automatic Data Processing equipment

AG/MET   — Agricultural/Meteorological Bulletin

ARPA   — Advanced Research Projects Agency

ARPANET   — Advanced Research Projects Agency Network

ASI   — Advanced Study Institutes Program of NATO

BASIS   — Battelle Automated Search Information System

BCD   — Binary Coded Decimal (code for representation of alphanumeric characters)

BODC   — British Oceanographic Data Center

BT's   — Bathythermographs

CAI   — Computer-Aided Instruction

CALCOMP   — California Computer Corporation

CDC   — Control Data Corporation

CHARNAL   — Character Analysis program of ENVIR

CHESS   — A Study by EPA on the human effects of sulfur dioxide

CRT's   — Cathode Ray Tubes

ECHKON   — Computer program for contouring meteorological data

EDGES   — Research program of the Chesapeake Research Consortium

EDMPAS   — Environment Dependent Management Process Automation And Simulation

EDP   — Electronic Data Processing

EDS   — Environmental Data Service of NOAA

EEC   — European Economic Community

ENDEX   — Environmental Data Index (part of EDS)

ENVIR   — Environment Information Retrieval program of GURC

ENVIRON   — Environmental Information Retrieval On-Line (EPA text oriented
               system)

EPA   — Environmental Protection Agency of the USA

FIC   — Federal Industrial Commission of the USA

FORTRAN   — Formula Translation (a computer language)

GARP   — Global Atmospheric Research Program of the WMO

GATE   — GARP Atlantic Tropical Experiment

GEMS   — Global Environmental Monitoring System of the U.N.

GEOSECS   — Geochemical Sections (A National Science Foundation oceano-
               graphic research program)

GIS   — General Information System (developed by IBM)

GOES   — Geostationary Operational Environmental Satellite

GRAPH   — A program module for presenting data in the ENVIR system.

GTS   — Global Telecommunications System

GURC   — Gulf Universities Research Consortium

GYPSY   — Information retrieval program

HASP   — Houston Automatic Spooling Program

HRIR   — High Resolution InfraRed photographs

IBP   — International Biological Program of the National Science Founda-
               tion

ICSU   — International Council of Scientific Unions

IDOE   — International Decade of Ocean Exploration of the NSF.

IGY   — International Geophysical Year

INFO 360   — General information management system developed by John Hopkins University Applied Physics Laboratory for IBM 360 computer

IO   — Input Output

IOC   — Intergovernmental Oceanographic Commission

IODE   — International Oceanographic Data Exchange (IOC Working Committee)

IRS   — International Referral Service of the UNEP

ITE   — Institute of Terrestrial Ecology at Merlewood, England

JOIDES   — Joint Oceanographic Institutions for Deep Earth Sampling

MEDLARS   — Medical Literature and Retrieval System

MEDLINE   — MEDLARS On-Line

MIS   — Management Information Systems

MONITOR   — Part of the TWDB computer system

MPNE   — Ministry for the Protection of Nature and the Environment of France

NATO   — North Atlantic Treaty Organization

NERL   — National Ecological Research Laboratory of the EPA

NOAA   — National Oceanic & Atmospheric Administration of the USA

NODC   — National Oceanographic Data Center of NOAA

NODC's   — National Oceanographic Data Centers (U.S., FRG, U.K., France, Spain, etc.)

NRIS   — Natural Resources Information System of the State of Texas

NSF   — National Science Foundation of the USA

PANDEX   — Current index to scientific technical literature by Lockheed

ROSCOP   — Report of Observations/Samples Collected by Oceanographic Programs

SACM   — Automatic contouring package of ACI's

SDS   — Scientific Data Systems, Inc.

STORET   — Storage and Retrieval system operated by EPA for water quality data

SYMAP   — Symbol Map (a program for graphic display)

System 2000   — A general information management system by MRI Systems Corp., Austin, Texas

TG4DA   — Topographically Guided 4-Dimensional Analysis (a computer program of GURC's)

TRANSDEX   — Translations Index (U.S. Joint Publications Research Service Translations)

TWDB   — Texas Water Development Board, State of Texas

TWODB   — Texas Water Oriented Data Bank of the TWDB

TYMNET   — A computer network operated by Tymshare, Inc.

UN   — United Nations

UNEP   — United Nations Environment Program

UNESCO   — United Nations Educational, Scientific & Cultural Organization

UNISIST   — UNESCO/ICSU World Science Information System

UNIVAC   — Sperry Rand Corp. trademark for computers

WODPS   — Water Oriented Data Programs Section of the Interagency Council on Natural Resources of the State of Texas

WMO   — World Meteorological Organization

REFERENCES

Anon. 1974 The monitoring of the environment in the United Kingdom: A Report by the Central Unit on Environmental Pollution. Pollution Paper No. 1. Her Majesty's Stationery Office, London. 66 pp. (ISBN O 11 750719 9).

Anon. 1975 Controlling Polution: A Review of Government Action Related to Recommendations by the Royal Commission on Environmental Pollution. Pollution Paper No. 4. Her Majesty's Stationery Office, London. 31 pp. (ISBN 0 11·659788 3).

Anway, J. D., Brittain, E. G., Hunt, H. W., Innis, G. S., Parton, W. J., Rodell, C. F., and R. H. Saver. 1972. ELM Version 1.0. 285p. Tech. Repr. No. 156, U.S. IBP Grassland Biome, Natural Resources Ecology Lab., Colorado State University, Fort Collins, Colorado.

Bergthorsson, P. and B. R. Doos. 1955. Numerical weather map analysis. *Tellus,* Vol. 7, No. 3, Aug. 1955.

Boltzmann, Ludwig. 1896. Vorlesungen uber gastheorie, I. Teil. J. A. Barth, Leipzig. 60 pp.

Boltzmann, Ludwig. 1898. Vorlesungen uber gastheorie, II. Teil. J. A.˙ Barth, Leipzig. p. 219-221.

Centres for Environmental Records. 1973. Vaughan Papers in Adult Education, No. 18, Univ. of Leicester, Department of Adult Education. September, 1973.

Cressman, G. P. 1959. An operational objective analysis system. *Monthly Weather Review*, Vol. 87, No. 10, Oct., 1959.

Jeffery, K. G. et al. 1974. The G-Exec system. Internal Report, Computer Unit, Institute of Geological Sciences, London.

Leopold, L. B., Clarke, F. E., Hanshaw, B. B., and J. R. Balsley. 1971. A procedure for evaluating environmental impact. Geological Survey Circular 645, U. S. Dept. of Interior, Washington, D.C. 13 pp.

Morisita, M. 1959. Measuring of interspecific association and similarity between communities. Memoirs of the Faculty of Science, Kyusha University, Series E, 3: 65-80.

NATO Scientific Affairs Division. 1967. NATO and Science-Facts about the activities of the Science Committee of the North Atlantic Treaty Organization 1959-1966. NATO Scientific Affairs Division, Brussels. 142 pp.

Odum, H. T. 1971. Environment, Power and Society. John Wiley and Sons, New York. 331 pp.

Ono, Y. 1961. An ecological study of the Brachuran community on Tomioka Bay, Amakusa, Kyusha. *Records of Oceanographic Works in Japan,* Spec. No. 5: 199-210.

Panofsky, H. A. 1949, Objective weather map analysis. *J. Meteor.* Vol. 6, No. 6, Dec., 1949.

Peachey, J. E. 1974. Environmental information and data handling. Roy Soc. (London), Proc., B. 185: 209-219.

Rannestad, A. (ed.). 1975. NATO Science Committee Year Book 1972 & 1973. NATO Scientific Affairs Division, Brussels. 362 pp.

Report of the Committee of Three. 1956. Non-Military Cooperation in NATO. NATO Information Service, B-1110 Brussells.

Rosenfeld, M. A. 1969. Marine data management—the views of university scientists: Proceedings of International Marine Information Symposium. Marine Technol. Soc., Washington, D.C. p. 105-123.

Shannon, C. E. 1948. A mathematical theory of communications. Bell System Tech. J., 27: 379-423, 623-656.

Watt, K. E. F. 1973. Principles of Environmental Science. McGraw-Hill Book Company, New York. 319 pp.

INDEX